.

WISSENSCHAFTLICHE BEITRÄGE AUS DEM TECTUM VERLAG

Unterreihe Wirtschaftswissenschaften

Band 4

Distributed Electricity Generation with Renewable Resources

Assessing the Economics of Photovoltaic Technologies in
Vertically Integrated and in Restructured Energy Markets

von

Raphael Edinger

Tectum Verlag
Marburg 1999

Die Deutsche Bibliothek - CIP-Einheitsaufnahme

Edinger, Raphael:
Distributed Electricity Generation with Renewable Resources.
Assessing the Economics of Photovoltaic Technologies in Vertically Integrated and in
Restructured Energy Markets.
/ von Raphael Edinger
- Marburg : Tectum Verlag, 1999
Zugl: Univ. Diss Hohenheim 1998
ISBN 3-8288- 8010-X

© Tectum Verlag

Tectum Verlag
Marburg 1999

CONTENTS

PART X. APPENDIX

ACKNOWLEDGMENTS

T his doctoral thesis is but the essence of a long research project and hardly capable of representing the wealth of information I learned and experiences I made both in Germany and in the USA. Before outlining the most crucial results of my doctoral project, I would like to acknowledge the support of people who have helped to realize my research project. This work could not have been done without the advice and assistance of many individuals and institutions in regard of both organizational and academic help as well as friendship.

It is Prof. Dr. Gerhard Scherhorn's merit to have me acquainted with the world of ecological economics. Attending his courses at Hohenheim University has profoundly influenced my understanding of the interdependences between economic and social / environmental phenomena. He is one of the very few scientists who are sensitive for the roots and restrictions of modern economic theory and the importance of understanding economics as a subsystem of our social and environmental world. Prof. Scherhorn has guided me through this research project and enriched my work with his counsel.

A European Recovery Program (ERP) scholarship of the German National Scholarship Foundation (Studienstiftung des Deutschen Volkes) and the German Ministry of Trade and Commerce financially supported my research year in the USA. I am especially grateful to Dr. Rupert Antes, Studienstiftung Bonn, who has provided encouragement and non-bureaucratic help on any occasion. Dr. Antes has helped me in organizing and financing my participation in academic and industrial conferences such as the USAEE/IAEE Conference on Energy Deregulation in Boston, October 1996, and the ASES/ISES Solar Energy Forum in Washington, D.C., April 1997. Dr. Antes has continuously accompanied my education at Hohenheim University and helped me with the organization of several language courses abroad as well as any other issue related to academics.

In Boston, I was rather unexpectedly introduced into the world of computer-aided dynamic modeling of real world phenomena. Prof. Matthias Ruth at the Center for Energy and Environmental Studies, Boston University, had invited me for this research year without any hesitation and helped me in both academic and personal questions and issues. Prof. Ruth allowed me to attend his class of STELLA dynamic modeling and continuously provided constructive input for my own doctoral project. While Prof. Ruth is one of the most hard-working people I have ever met, he took his time for me any time I needed assistance.

In regard of visa issues, I am especially grateful to Mrs. Ruth Sieck. Since I entered the US on an erroneously issued student visa instead of a research visa, Mrs. Sieck provided the most valuable support in changing my visa status and helping me with the US immigration offices. Mrs. Sieck also helped me to find accommodation when I was desperately looking for a place to stay being late for the US semester. With the International Fellowship House in Boston, Marlborough Street, I found not only a home, but also a family of some 20 international students living together forming a community far away from home.

In Washington, I met Prof. Sanjay Kaul, Fitchburg State College, who initiated a research and publication project. Meanwhile, Prof. Sanjay and his family have become good friends of mine. He encouraged me to soon start writing scientific articles and books. With his wife and his children, I have learned to know a loving Indian family who acquainted me with the Indian culture and the extremely spicy dishes of Kashmir.

In course of the ERP scholarship program, I spent several weeks at the Rocky Mountain Institute (RMI), Snowmass, Colorado. At RMI, I worked with Richard Pinkham and Amory Lovins on Distributed Water and Power Systems. Amory Lovins has impressed me not only with his bright and unconventional mind, but also with his being even more restless and hyperactive than myself. Since Amory was mostly on the road, I usually worked together with Richard Pinkham who was a most supportive supervisor and helped me understand the importance of ecologically sustainable water systems. Mr. Pinkham also gave me sufficient free time to work on my own and proceed with my doctoral research program.

During my second year of research, I visited several research institutions such as the Wuppertal Institute for Climate, Energy, and the Environment, Wuppertal, Germany, and the Fraunhofer Institute for Solar Energy Systems. Cooperation with the local BUND Group, Nürtingen, has improved my understanding for implementing renewable technologies into real projects and broadened my knowledge on climate change issues. I have to thank Mr. Otmar Braune for his helpful advice and insightful information as well as for his support with information material and specific literature.

Since the Institute for Consumer Economics at Hohenheim University is about to be closed, I am both proud and sad to be one of the last doctoral students at this institute with its many bright and most helpful people. The Institute for Consumer Economics is one of the farsighted academic economic centers having performed valuable research and teaching in the field of sustainable energies and ecological economics.

Raphael Edinger
Hohenheim University, Germany

PART I.

INTRODUCTION

"Using high-quality electric power for purposes such as cooking and space heating is like cutting butter with a chainsaw."

Amory B. Lovins, Rocky Mountain Institute

1. Preface

Energy is necessary to provide services to our daily lives. Solar radiation empowers nature to bring forth life as well as natural resources that can be used by human beings. Today's technology uses energy much more efficiently than a century or even only decades ago. A growing worldwide population and global development, however, urge us to utilize limited fossil fuels even more carefully. The depletion of natural resources and global climate change are a challenge for the living and future generations.

Electricity is one of the highest quality forms of energy currently available. Due to its versatility, electric power is used for a broad array of appliances. Comparatively cheap electric power generation has lead to the universal use of electricity. However, electric power is usually being generated by burning fossil fuels and implies a multitude of energetic losses from electric power generation to final electric power consumption. Today, we have to critically rethink the universal use of electricity and come forth with solutions to employ our planet's energy sources more reasonably.

It is not enough to further increase the energy efficiency of electric power generation and electric end-use appliances. In many occasions, it is preferable to switch the energy source. As Amory B. Lovins critically indicates, it is not sensible to first burn fossil fuels to make steam turning the rotors of a turbine, which generates electric current that is then transferred over long distances to the end-user, who finally converts this high-quality electric power back into heat for cooking or space heating. It is easy to understand that from an energetic point of view, this cannot be very efficient with electric losses occurring at many stages of the transformation and delivering process.

But this may well be and often is the most cost-effective way to produce consumer services. Low energy prices and the centralized monopolistic structure of the electric power industry have made electric power a vital energy form used by both industrial and residential consumers.

A rather new, but elegant way to produce electric power is photovoltaic technologies (PV). PV directly converts solar radiation into electricity without any mechanically moving parts or the burning of fossil fuels. PV has provided reliable power to space applications and residential houses situated in remote locations far off the central grid system. The residents of remote PV powered homes value the electric power they generate much higher than grid-connected customers. Since PV electricity is expensive, it is used for those appliances only that could not be powered effectively otherwise (e.g. radio, lighting, telephone, computers, etc.). Heating and cooling services can be provided more efficiently by natural gas powered appliances or solar water heaters. Cutting down the load side first (insulation of the house; use of energy-efficient appliances; etc.) is both energy-efficient and economic at these residencies.

Grid-connected electric customers have benefited from comparatively low prices and the easy-to-use universality of electric power. We therefore use this high-quality energy form much more generously than necessary at today's state-of-the-art technologies. Electric power generation with renewable resources has been neglected for its higher costs than conventional power generation with fossil fuel technologies. However, in order to use our energy resources efficiently and in an environmentally and socially sustainable way, we have to rethink our power basis and carefully assess the real costs and benefits of new forms of energy technologies. In the context of this research work, we will therefore examine the economics of small grid-connected residential photovoltaic systems.

2. Deriving the Issue of Photovoltaic Power Generation

2.1 Energy Sources and Uses

Photovoltaics are not the only form of renewable power generation. It is helpful to initially outline a systematics of energy sources and uses.

Figure 1-1. Systematics of Energy Uses

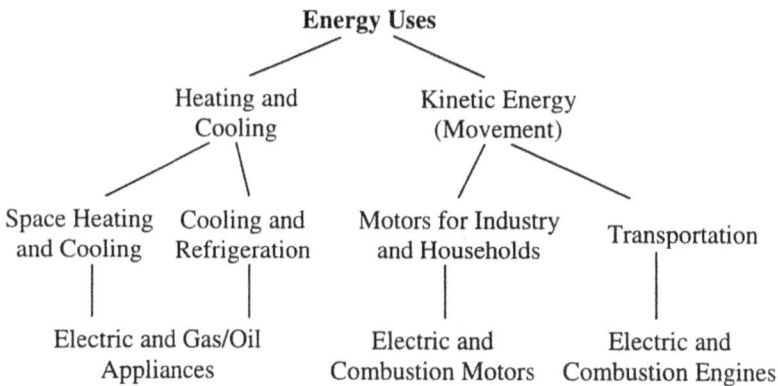

We use energy for a broad range of purposes (see Figure 1-1.). Heating and cooling are one major application. These services can be done by either electric or gas/oil water heaters and stoves or refrigerators. Kinetic energy is necessary for driving machines and transportation purposes. Electric and combustion motors and engines power household appliances, industrial machinery, and vehicles. As we see in this outline, the final services can be provided by either electric or directly fossil fuel powered machines.

Fossil fuels are today's ubiquitous power source. Burning coal, natural gas, and oil delivers reliable energy services at almost any location. Nuclear power is another form of fossil fuels, yet requires a distinct transformation and generation process.

For centuries, humankind's primary energy basis was renewable resources such as solar, biomass, hydro, and wind energy. Windmills were used to grind corn and cut logs; hydropower stations produced mechanical and electric energy; wood logs provided space heating and cooking energy; and solar radiation was used for drying food and clothing. Solar energy is the principal power basis for any other energy form. The universality and low-cost of fossil fuel technologies, however, put an end to the predominance of renewable power usage in the course of the industrial revolution.

Figure 1-2. Systematics of Energy Sources

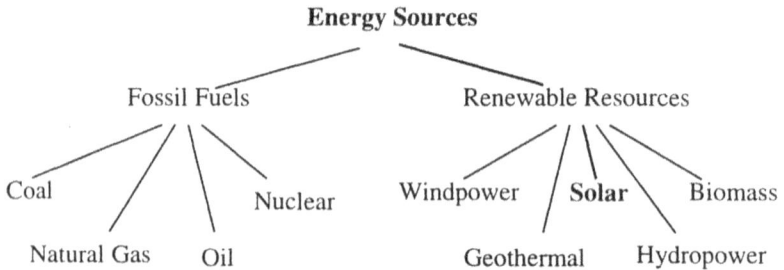

Energy Sources

Fossil Fuels   Renewable Resources

Coal   Nuclear   Windpower   Solar   Biomass

Natural Gas   Oil   Geothermal   Hydropower

Today, new state-of-the-art technologies have been developed to efficiently and cost-effectively make use of renewable energy sources (Figure 1-2.). Modern wind machines are one of the most cost-efficient ways of renewable electric power generation. In Germany, wind power machines of the 600-kWp class sited in wind speed areas of an annual average of 6.3 m/s in 30 meters above ground level, operate at 9 cents per kilowatthour.[1] Medium size and microhydro power generation can produce even cheaper electricity, although the economics are highly dependent on local conditions. Hydropower is constrained to locally available water resources and struggles with environmental problems such as handicapping fish migration and fertile sediment transportation through floating waters.

Biomass and geothermal are also promising renewable power technologies and could effectively contribute to a more sustainable energy system. In the context of this research work, we will focus on solar energy and examine the economics of solar electric power generation with photovoltaic technologies.

2.2 Solar Energy and Photovoltaic Power Generation

The rich variety of solar technologies sometimes leads to confounding discussions on the economic and technical feasibility of solar energy. Figure 1-3. tries to systematically derive the branch of solar electric power generation we will assess in this research work.

First of all, *solar electric* power generation has to be distinguished from *solar thermal* technologies used for water and space heating as well as cooking purposes. Solar thermal collectors directly convert solar radiation into heat, not into electric power.

For solar electric power generation, two basic technologies are currently being used. *Solar thermal electric* systems basically concentrate the solar rays with reflector lenses onto a focus point where a liquid is heated and vaporized. The produced steam then turns a turbine, generating electric power. Central receiver systems, such as the famous US American Solar I and Solar II power stations, are large central power stations located in desert areas. These systems use an array of optical mirrors to focus sunlight onto a tower-mounted central receiver filled with a fluid that produces steam, driving a turbine to produce electricity. Parabolic trough systems concentrate solar radiation onto a fluid-filled pipe running the length of the trough. The Dish-Stirling system uses a small dish-shaped reflector with dual-axis tracking to heat a fluid powering a small engine mounted at the focal point of the dish.[2]

Photovoltaic (PV) technologies do not rely on steam-powered turbines or other mechanical devices. Photovoltaic technologies use a semiconducting material to convert sunlight directly into electricity. PV systems can either be integrated into

---

[1] Molly, Jens Peter, and Knud Rehfeldt (1997). *Wirtschaftlichkeit von Windanlagen / Aktuelle Kostenentwicklung*, p. 7.

[2] Energy Information Administration EIA / US Department of Energy DOE (1995). *Renewable Energy Annual 1995*, p. 101.

the existing electric grid system or provide electric power to remote locations in terrestrial and space applications. Grid-connected PV systems can be installed in large units like conventional central power stations or be mounted in small modular units (distributed power generation) at grid locations where electricity is the most valuable and needed. Distributed PV systems can support the utility grid system or be operated at the location of the residential customer.

Figure 1-3. Systematics of Solar Energy

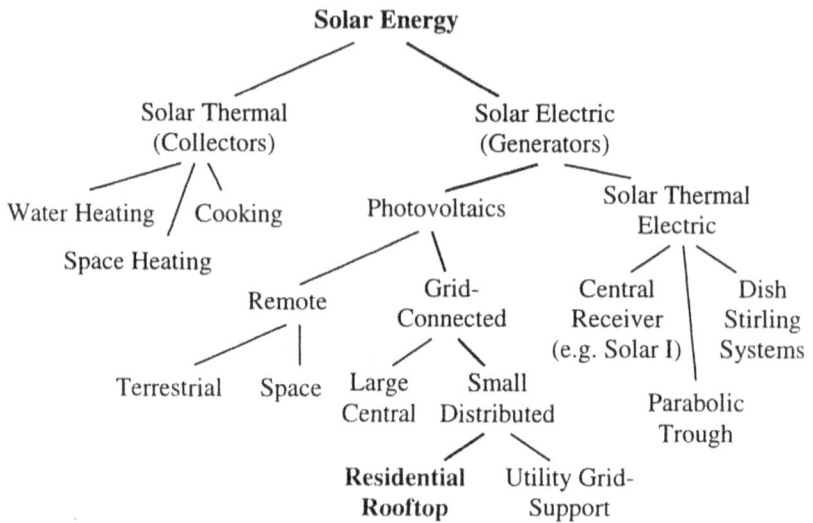

For this research project, we will examine the economics of electric power generation with small grid-connected residential PV sytems installed at the customers' rooftops. Several reasons contributed to choosing this power technology for detailed assessment:

Technology

➤ The technology for grid-connected rooftop PV systems is currently available and has proved its reliability in a broad array of applications.

➤ Solar radiation is available in most locations and can be used for electric power generation in developing and industrialized countries.

➤ Photovoltaics can be installed at customer sites and added incrementally due to their modular and small-scale character.

Cost

➤ Costs for grid-connected PV systems have declined significantly over the last decades. Further significant cost-reductions are expected due to new technologies (thin-film materials) and beginning mass production.

➤ For rooftop PV systems, no additional location area has to be purchased and maintained. The systems are just mounted on existing residential rooftops.

Market

➤ The market for residential PV sytems is potentially large and prospective.

➤ Surveys and utility programs indicate a wide public acceptance for photovoltaic technologies, as will be discussed below in more detail.

Environment

➤ PV power generation does not produce any toxic emissions. Most commonly used PV cell materials can be recycled and are non-toxic.

➤ The energy payback-time for PV cells and systems has decreased to several months (dependent on system features) and is within the system lifetime.

➤ Grid-connected PV systems do not need expensive and environmentally problematic battery storage.

➤ Located near the load side, PV power generation features low energy losses in terms of energy transmission and distribution.

In spite of their versatility and large potential contribution to our current electric power system, photovoltaic technologies will not be able to provide the total energy necessary for industrial societies. But this is not necessary either. From an energetic point of view it is generally advisable to cut down the customer's load side first before trying to provide the remaining energy demand with the most environmentally friendly and economic power option available. It is important to rethink the universal use of electric power in our society. As we have seen in Amory Lovin's quotation at the opening of this chapter, it is not sensible to transform high-quality electric power (even environmentally-friendly solar electric power) back into heat. Next to photovoltaic technologies, we will therefore have to rely on solar thermal technologies that cost-effectively convert solar radiation directly into heat.

But even our electricity needs cannot be provided by photovoltaics alone. This is due to the fact that sunshine does not always coincide with electric power demand in terms of time and volume. Energy storage is necessary, but always implies energetic losses. A promising technology is hydrogen storage and hydrogen fuel cells, although this technology is not yet available for household applications. While photovoltaics may not be able to cost-effectively provide the total electricity needs, a multitude of renewable power technologies such as small-scale hydropower, wind power, photovoltaics, biomass, geothermal and others can build our future electric power basis.

Renewable technologies are capable to provide the necessary energy services to residential customers. Yet, it may be difficult to transfer this concept to the industrial sector. The cement and steel industry, for example, rely on high energetic resources to transform virgin materials into usable endproducts. Renewable technologies are usually not available in the necessary intensity and quantity. It will be difficult to reach the required temperatures for steel production with biomass and wood firing. It has yet to be examined whether new production methods and new materials can cut down the industry's energy needs to be effectively provided by renewable energies. Regarding energy sources and energy uses in developing countries, however, also indicates that industrialized countries generally live on a consumption level hardly compatible for sustainable economies. We may therefore have to change our patterns of industrial production and private consumption as well as cutting back the use of energy-intensive materials by our society.

3. Outline of the Doctoral Thesis

In this doctoral thesis, we will examine the economics of grid-connected photovoltaic power systems on the rooftops of residential customers. The assessment will refer to US American and German data and case studies. Monetary values are used for the specific case (US$ and cents for the US; DM and DPf for Germany). The calculations are based on an exchange rate of 1.80 DM/US$.

Part II outlines the research target and presents the hypotheses and assessment tools. Part III gives an overview of the US American and the German restructuring process and discusses historic aspects, crucial legislative issues, as well as new institutions and entities in a liberalized electric power market.

In Part IV, photovoltaic technologies are presented and explained in their systematic context. Types of PV cells, PV modules, and PV systems are explained and characterized by their technical features and current as well as expected costs.

Part V introduces economic aspects of photovoltaic power generation in a regulated and vertically integrated electricity market. This part contains a short and basic comparison of photovoltaic and fossil fuel power technologies and addresses subsidy and taxation issues.

PV power production in a restructured and liberalized market environment is subject of Part VI. This part outlines the new market structure and crucial investment decision parameters for electric utilities. Viable strategies for photovoltaic power generation in a liberalized energy market are presented with reference to current US utility projects.

Part VII exposes the concept of distributed power generation and the creation of a decentralized electricity system based on renewable technologies. Both the environment and electric utilities can benefit from distributed power options integrated into the existing grid or implemented into electrically isolated micro-grid systems.

After this theoretical and empirical examination of photovoltaic technologies and electric restructuring, the economics of PV power generation is carefully assessed in Part VIII. The analysis is performed with the new STELLA dynamic modeling software, integrating monetary aspects and market conditions and experimentally varying the economic situation of PV technologies. In course of this assessment, a strategy for PV power generation in liberalized power markets will be developed and finally discussed in Part IX.

The Appendix in Part X includes an outline of units of measure; an extensive glossary for technical terms and their English-German translation; a list of references; and an index of tables and figures.

PART II.

RESEARCH TARGET, HYPOTHESES, AND ASSESSMENT TOOLS

4. Target of Research

A large quantity of publications in the photovoltaic (PV) power market focuses on the technical advances of PV systems. Only a few studies are available that analyze economic aspects of photovoltaic energy generation. Electric utilities as well as academic research has concentrated on technological issues since PV appeared to be far from being a cost-competitive electric power option in today's energy market.

Nevertheless, there are some utilities that have already begun to use PV technologies to generate electric power for the central grid. While the majority of utilities assess PV primarily for further research studies only, there are proactive utilities that regard PV technologies to be already cost-competitive in their business today.

Grid-connected PV systems appear to be the most competitive PV technology in electricity markets of industrialized countries that do already have central power grids. Installations on the roofs of electric customers have technical and economic advantages compared to large centralized PV generation.

This research work examines the economics of grid-connected rooftop photovoltaic systems in regulated and restructured electricity markets. Data from the USA and Germany is analyzed to assess the economics of PV power generation in various market scenarios.

5. Hypotheses

Electric utilities have operated in a regulated market of vertically integrated economic entities. Restructuring is about to drastically modify the appearance of the electric power market. Competition and emerging new technologies challenge the monopolistic utility market. Electric restructuring has just begun in the United States and is about to influence energy policies in Europe and Germany, too. The current and future market structure in the electricity industry changes the economics of small-scale, renewable power generation technologies such as PV.

"Most of the U.S. utilities really do not know what they do", reasoned a manager of the Sacramento Municipal Utility District at a U.S. energy congress in 1997.[3] Utilities probably underestimate the role of electric customers in a future restructured electricity market. Private consumers will not only have the choice of a broad array of electric services and the right to choose their power supplier. For the first time in the history of regulated utilities, private consumers do also have the

---

[3] Personal Conversation with Daniel D. Whitney, SMUD, at the ASES/ISES Solar Energy Forum, Washington D.C., April 22nd, 1997.

opportunity to produce their own current and sell it to others or even back into the central grid. Photovoltaics is but one of several small-scale power technologies that allow small power producers (SPPs) to compete against large electric utilities. The role of private electric customers changes from consumers only to potential power producers and could bring forth serious competition for traditional electric utilities.

Based on these ideas, the following hypotheses are to assess the economics of PV power generation.

Hypothesis 1. The Economics of PV Power Generation in a Regulated and Vertically Integrated Electricity Market

1.1 In a regulated and vertically integrated market, solar electricity generation with small, grid-connected photovoltaic systems is not a cost-effective alternative for electric utilities.

1.2 In a regulated and vertically integrated market, solar electricity generation with small, grid-connected photovoltaic systems is not a cost-effective option for residential consumers either.

Hypothesis 2. The Economics of PV Power Generation in a Restructured and Liberalized Market Environment

2.1 In a deregulated and liberal market, photovoltaic electricity generation will become a viable power technology for electric utilities.

2.2 In a deregulated and liberal market, electric utilities will face new competition from consumers producing their own photovoltaic electricity and selling it to the grid.

Hypothesis 3. Distributed Power Generation: Creating a Decentralized Electricity System

3.1 Electric restructuring and technological advancement will for economic reasons create a rather decentralized (distributed) electric power system.

3.2 Small, grid-connected photovoltaic systems on residential rooftops will be a cost-effective part of distributed power systems.

6. Assessment Tools

6.1 Modeling Complex and Dynamic Real-World Systems

In order to analyze and solve real-world problems, we usually have to break down the general problem into various sub-systems to reduce complexity. Of course, we should always bear in mind the holistic structure of the original question.

There are numerous analytical tools to examine complex problems. Some of these tools are descriptive, others predictive; some analyses use the calculation speed of modern computers to cope with a large amount of data and complex information.

One method to assess empirical questions is to set up models and experimentally modify selected variables. Skeptics argue that models do not represent real-world systems but create an artificial analysis environment. However, models do have important advantages for solving complex problems. Matthias Ruth, Boston University, and Bruce Hannon, University of Illinois, identified four purposes of creating analytical models:

> "First, models enable you to experiment: A good model of a system lets you alter its components and experience the effect of such changes on the system. Second, good models give you insight into the future course of a dynamic system. Third, good models lead you to further questions about a system, what underlies its behavior, and how applicable the principles discovered in the modeling process are to other systems. Fourth, good models are good thought-organizing devices."[4]

This doctoral thesis is trying to comply with these targets of complex and dynamic modeling. The analysis will be based on the STELLA dynamic modeling software as described in the next paragraph. The models developed will be easy to understand and can be implemented universally. Models are used to solve complex financial calculations and assess system behavior in a changing market environment. Forecasting models simulate the starting-point and impacts of mass-production and expected cost-decline for photovoltaic technologies. All models are based on the most currently available data in order to reach the highest possible level of empirical validity. The model character of the examination also allows updating variable values whenever new data is available. The models' assumptions can be easily verified and modified with data from political legislation, industrial companies, or private residential customers.

---

[4] Ruth, Matthias and Bruce Hannon (1997). *Modeling Dynamic Economic Systems*, p. 21.

6.2 STELLA Dynamic Modeling Software

Electric utilities traditionally base their investment decisions on conventional cost calculation techniques such as levelized cost calculations. The changing structure of the electric power market, however, features a large array of variables that have to be assessed for future investment decision making. These variables are highly interdependent. Conventional cost-calculation techniques are not able to handle the feedback processes necessary for examining the cost and benefits of various power generation technologies in a changing market environment.

The STELLA Dynamic Modeling Software was developed to model dynamic economic and environmental systems. Feedback processes can easily be visualized through this intelligible Windows-based programming tool. The interactions of cost factors, market scenarios, and technological progress can be modeled and analyzed with the STELLA software.

Sensitivity analyses permit to assess the influence of input variables on a complex dynamic system. The economic models give predictions of the short- and long-term outcomes of proposed actions. STELLA effectively combines mathematical models with experimentation. Part VIII will explain the modeling process in detail and carefully introduce into developing complex economic models.

PART III.

ELECTRIC RESTRUCTURING OVERVIEW

7. Electric Restructuring in the USA

7.1 The Rise of Vertically Integrated Electricity Monopolies

Thomas A. Edison was the American pioneer of electric power generation. Not only did he develop the first incandescent lamp that made him famous all over the world, he also designed the first electric generation station in the United States. On September 4, 1882, Edison's Pearl Street Station opened in a New York neighborhood. The district included 1500 gas light customers and the potential for 750 electric motors. The Pearl Street power station could sell electricity both day and night and started by providing power for 85 customers with 400 lamps.[5]

In 1896, George Westinghouse developed the large hydroelectric power system at Niagara Falls. Meanwhile the issue of long-distance power transmission was technically solved. The Niagara power station set a new dimension of central power generation and distributed its power to a multitude of electric customers. With the installation of a large central power grid, electric power stations spread rapidly throughout the United States. Electric power became an important energy source for private households and the industry.

In the first decades of the electric power industry, average cost of electric power production declined with the number of customers served and the size of the power plants increasing. These economies of scale made it cost-effective to install larger central power plants (see Figure 3-1.). Larger steam turbines reduced average heat rates from 92,500 Btu per kWh in 1902 to as low as 20,700 Btu per kWh in 1932.[6]

In the first days, individual entrepreneurs such as Thomas Edison owned the electric power stations. With the size of power stations increasing, investor-owned companies provided the necessary capital for generating capacities and distribution lines.

Under the pressure of cost advantages through economies of scale, small private and municipal power companies merged with larger companies. Holding companies reduced the number of electric utilities substantially. In the late 1920s, the 16 largest electric power holding companies accounted for 75 percent of U.S. electric power generation.[7]

Declining energy demand during the Great Depression lead to utility stocks collapsing at Wall Street. The financial situation of many holding companies could not cope with the new economic situation of decreasing revenues. In this situation,

---

[5] Hyman, Leonard S. (1997). America's Electric Utilities. Past, Present and Future, pp. 84-85.

[6] Energy Information Administration / US Department of Energy (1996). *The Changing Structure of the Electric Power Industry: An Update,* Appendix A, p. 105.

[7] Ibid, p. 106.

the system of large holding companies proved unstable. The Securities and Exchange Commission and the Federal Trade Commission initiated investigations resulting in critical reports on the financial and economic practices of the holding companies. They were held responsible for improper accounting; manipulating subsidies; stock watering; and capital inflation.[8]

**Figure 3-1. Economies of Scale: Declining Average Cost (c/kWh)
due to Larger Quantity of Electricity Sold**

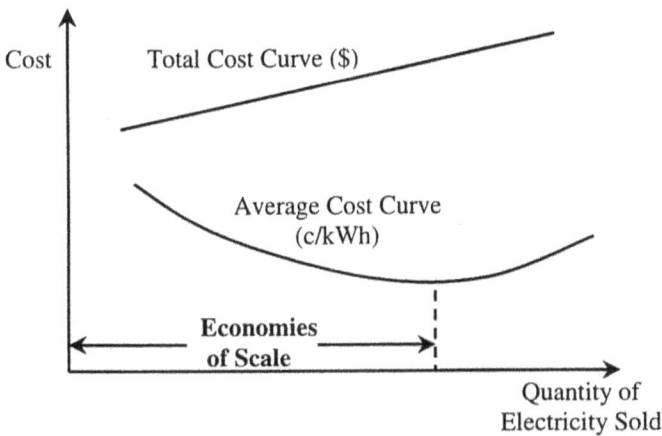

Edinger, Raphael and Sanjay Kaul (1999). *Renewable Resources for Electric Power. Prospects and Challenges.* Quorum Books, Westport/USA. In press.

The reports led to the first pieces of federal legislation concerning the restructuring of the electricity industry. In 1935, the Public Utility Holding Company Act (PUHCA, Public Law 74-333) was enacted to destroy the large trusts that controlled the U.S. electric and gas distribution industry. These holding companies often owned a multitude of electric utilities and were themselves owned by other holding companies. The complicated structure of these trust pyramids could reach

---

[8] Brennan, Timothy J., Karen L. Palmer, et al. (1996), *A Shock to the System. Restructuring America's Electricity Industry*, p. 23.

up to ten layers, with three huge holding companies controlling almost half of all electricity generated in the United States.[9]

PUHCA, however, did not address the issue of vertical integration. Electric utilities generated and transmitted electric power over long distances through the central grid, and finally distributed the electricity to households and industry. With the integration of the three functions power *generation*, long-distance *transmission*, and local *distribution* a system of vertically integrated utilities emerged, each serving its own service territory as a regionally monopolistic power provider.

In fact, vertical integration was not regarded to be problematic, since the electric power industry constantly decreased electricity rates to private and commercial customers. Technological advancement in fossil fuel power generation technologies as well as constantly growing demand for electricity after the Great Depression allowed utilities to decrease electric rates. While the number of electric customers quadrupled between 1925 and 1970, plant capacity increased thirteenfold and electric sales skyrocketed by a factor of 25. Between 1906 and 1970, the average residential rate dropped from nominal 10 cents to nominal 2.6 cents per kWh.[10]

Under these market circumstances, PUHCA seemed to be an effective instrument of legislation to control the electric industry. Low electric rates and growing demand was seen as an indicator of a well-functioning market. It was not until the 1970s that the structure of the U.S. electric power industry was again debated.

7.2 Milestones in Electric Restructuring

Crucial changes in the electric utility industry occurred after 1965, the year with the greatest rate reductions and highest electric utility stock prices. In the years following, capital expenditures rose for environmental reasons and technical problems with the construction and maintenance of large nuclear power stations. New generating facilities did not provide better heat rates but were much more expensive to set up than older power plants. Insufficient back-up capacity caused the Northeastern Blackout in 1965 through an equipment failure at Ontario Hydro and showed the unreliability of parts of the U.S. central grid system.[11]

In October 1973, the Arab nations answered the U.S. and Dutch support for Israel with an oil embargo. Crude oil prices quadrupled and industrialized nations became aware of the insecurity of international energy supply with fossil fuels.

---

[9] Energy Information Administration / US Department of Energy (1993). *The Changing Structure of the Electric Power Industry, 1970-1991*, p. 24.

[10] Fox-Penner, Peter (1997). *Electric Utility Restructuring. A Guide to the Competitive Era*, p. 12.

[11] Hyman, Leonard S. *America's Electric Utilities. Past, Present and Future*, pp. 135-136.

The rise of the U.S. nuclear power industry was severely hit by the accident at Three Mile Island in 1979. New nuclear power plant projects were cancelled or stopped. In Europe, the nuclear accident at Chernobyl/Ukraine in 1986 reinforced environmental and economic skepticism against nuclear power generation.

These events caused the U.S. government to undertake a new effort to restructure the monopolistic electric utility industry. In 1978, the Public Utility Regulatory Policies Act (PURPA) was passed to promote energy conservation and decrease U.S. dependence on international fossil fuel imports. PURPA was highly important for renewable energy advocates since it created a new class of nonutility generators, small power producers (SPPs) whose capacity did not exceed 80 megawatts. Utilities were required to purchase power from the SPPs at the utility's avoided cost, i.e. the cost for incremental additional capacity.[12] The avoided cost rate was substantially higher than the market price for electricity. Therefore, PURPA increased the number of non-utility electric power producers significantly. With SPPs primarily using renewable resources such as water and wind power, PURPA also initiated a more diversified resource mix for electric power. The Federal Energy Regulatory Commission (FERC) was to watch over the electric utilities and their compliance to PURPA.

The Energy Tax Act (ETA) of 1978 was passed in the same year as PURPA, also in response to the unstable energy markets of the 1970's. The ETA encouraged investment in cogeneration equipment and solar and wind. ETA allowed subsidizing and promoting renewable energies directly through taxation practices. However, the ETA tax incentives were cut with the tax reform legislation in the 1980's.

In 1963, the Clean Air Act (CAA), amended in 1967, 1970, 1977, and 1990, established a national regulatory system to protect and enhance the air quality. Electric utilities were held responsible for large shares of the air pollutants. The Clean Air Act restricted the emission of particulate matter, sulfur dioxide, nitrogen dioxide, carbon monoxide, ozone, and lead. The 1990 amendments to the CAA targeted ground level ozone, stratospheric ozone depletion, air toxins, and acid rain.[13]

The Energy Policy Act (EPA) of 1992 targeted increasing the share of nonutility power generation. EPA created a new category of electricity producer, the exempt wholesale generator, which enlarged PUHCA's restrictive definitions for nonutility

---

[12] Energy Information Administration / US Department of Energy (1997). *Renewable Energy Annual 1996*, p. 101.

[13] Energy Information Administration / U.S. Department of Energy (1995). *Electricity Generation and Environmental Externalities: Case Studies*, p. 17.

power generators. The law was the first to allow FERC to open up the national electricity transmission system to wholesale suppliers.[14]

7.3 Toward Liberal Power Markets

While all these legislative decisions instituted a restrictive framework for the electricity industry and cut back the power of monopolistic utilities, it was yet clear that the electric power sector had to be transformed into a more competitive market environment. The U.S. gas and telecommunications industries are examples of successfully liberalizing vertically integrated monopolies. Competition promises lower production costs and higher efficiencies, lower prices as well as high-quality customer service with electric consumers able to choose from a wide array of service providers.

In the first decades of the electric power industry, scientists and policymakers regarded the electric power system as a natural monopoly, where a single utility could provided electric service for a designated service territory at lower cost and thus lower rates than two or more competing providers. In the U.S. electric utility industry, high fix cost for generating facilities and distribution lines as well as economies of scale supported the natural monopoly view until the late 1970s.

Technological advancements in electric generation modified this fundamental belief in the 1980s and 1990s. Small and modular power technologies such as gas turbines, derived from the modern jet aircraft industry, as well as small renewable generating technologies such as micro-hydro and wind turbines generated electricity near the load side and created the idea of a more decentralized system. From an energetic perspective, combined-cycle gas turbines reach higher maximum efficiencies at a smaller scale and feature lower average cost than conventional central power plants.

Regulators finally decided to solve the natural monopoly problem with unbundling the integrated electric utilities into three separate parts, electric power *generation*, power *transmission*, and electric *distribution* (see Figure 3-2.). Transmission and distribution were still regarded to be natural monopolies, since investments in the central grid and distribution network required high capital expenditures. It did not seem reasonable to have several networks operating in parallel at the same time. Power generation, however, is no longer regarded a natural monopoly. Competing power generators are expected to achieve higher cost reductions than monopolistic firms serving their assigned territories do.

---

[14] Energy Information Administration / US Department of Energy (1996). *The Changing Structure of the Electric Power Industry: An Update,* p. 24.

Figure 3-2. Unbundling of Vertically Integrated Monopolies

Vertically Integrated Monopolist		Unbundled Electric Utility

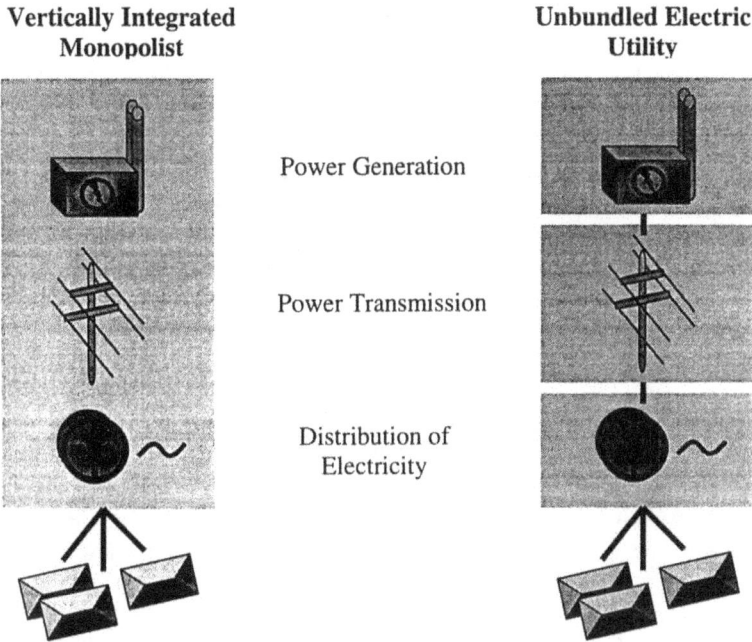

Power Generation

Power Transmission

Distribution of Electricity

Edinger, Raphael and Sanjay Kaul (1999). *Renewable Resources for Electric Power. Prospects and Challenges.* Quorum Books, Westport/USA. In press.

On April 24, 1996, The Federal Energy Regulatory Commission (FERC) issued Order 888 that encouraged wholesale competition and open access to the public utility grid system. FERC Order 889, issued at the same time, required utilities to establish electronic systems to exchange information about available transmission capacity. In consequence, several U.S. State laws gradually opened the market of power generation to new electric service providers and power marketers.

On September 23, 1996, California enacted Assembly Bill 1890. AB 1890 deregulated the California electric utility industry in several phases, beginning on January 1st, 1998, with a transition period of 1998 through 2001. The integrated energy services of each utility will be "unbundled". Independent entities within a utility have to be established to separately operate generation, transmission, and distribution. The level of competition depends on each of the three service domains. Regulated monopoly status is allowed for transmission and distribution only. Power generation will become competitive and operate under free-market principles.

While a multitude of service providers will compete for customers and contracts, the electricity price will be set by daily bids through a newly established Power Exchange (PX), or be negotiated in individual "direct access" contracts between individual buyers and sellers.

Utilities are required to transfer control, but not ownership, of their transmission lines to a central institution, called the Independent System Operator (ISO). The ISO will coordinate energy demand and supply and will grant or reject access to the transmission grid to power sellers on a non-discriminating basis. The California Public Utility Commission (CPUC) will guard over electric transmission and distribution.

7.4 California Pioneering the Deregulation Process

California has comparatively high electricity rates, which may have been one reason for the State pioneering electric restructuring. California opened its electricity markets to competition on March 31st, 1998, after a 3-month delay through computer problems. Ten million customers of Southern California Edison Company (SCE), Pacific Gas & Electric (PG&E), and San Diego Gas & Electric Company (SDG&E) could now choose their electric power provider. Consumers rather reluctantly made use of their right of free choice, with fewer than 40,000 of the 10 million eligible customers switching to new power providers as of May 1998. Approximately 10% of California's large commercial and industrial customers switched their electricity provider.[15]

Customers of California municipal utilities are currently exempt from the deregulated market system. Municipals will have the option to join the system in 2 years.

8. Electric Restructuring in Germany

8.1 The Origins and Development of the German Electric Power Industry

In the early1880s, Rathenau and Siemens founded the German Edison Society for Applied Electro-Technology (DEG). Their privately owned enterprise served a club and later a residential street with electric light. The first publicly owned power

---

[15] Energy Information Administration EIA / US Department of Energy DOE (1998). *Industry Developments: California Successfully Starts Deregulated Electricity Market*, p. 9.

stations were started in Stuttgart (1882) and Berlin (1884), but the privately owned utilities became dominant since they were more willing to take higher risks in the newly established electric power market.[16]

In the 1920s, the German electricity system concentrated through mergers and acquisitions like the U.S. American electric industry. As the concentration process increased the size and influence of single utilities such as the Rheinisch-Westfälische Elektrizitätswerke (RWE) and communities started to fear the economic power of large electric suppliers, the German government intervened with a demarcation agreement. The electric utilities were granted the right to exclusively serve a geographic area of electric customers and had to pay a concession fee to the cities or communities. This agreement between politicians and electric utilities initiated the monopolistic structure of vertically integrated electricity companies.

The large utilities also operated the high-voltage transmission grid and today form the Deutsche Verbundgesellschaft (DVG). In the Old States, Preussen Elektra AG, Hamburgische Electricitäts-Werke (HEW) AG, Vereinigte Elektrizitätswerke Westfalen (VEW) AG, RWE Energie AG, Bayernwerk (VIAG) AG, Berliner Kraft- und Licht AG (BEWAG) and the Energie Baden-Württemberg (EBW) AG are the dominant investor-owned electric utilities. These companies are highly interconnected with capital links, as shown in Figure 3-3. They also own shares of the Vereinigte Elektrizitätswerke AG (VEAG) in the New States.

The German electricity sector has become a highly integrated and monopolistic power market. Between the 1930s and today, the number of public electric utilities declined from more than 16,000 to only 900. The large DVG companies account for 75% of today's installed generating capacity. Apart from investor owned utilities, private enterprises, regional utilities, and others, approximately 570 municipals are currently serving electric customers in cities and communities. Some of them produce electric power and purchase necessary surplus-electricity from the large DVG-companies, others just trade electricity. In 1993, the DVG companies held shares of between 10 and 99% in 57 of the municipals.[17]

The concession fees of electric utilities considerably add to the financial budgets of cities and communities. In turn, the communities also function as regulators by extending concession agreements and by controlling electricity tariffs to some degree.

---

[16] Müller, Jürgen and Konrad Stahl (1996). *Regulation of the Market for Electricity in the Federal Republic of Germany*, pp. 277-284.

[17] Mez, Lutz (1997). *The German Electricity Reform Attempts: Reforming Co-optive Networks*, pp. 232-235.

Figure 3-3. Capital Links between German Electric Utility Companies

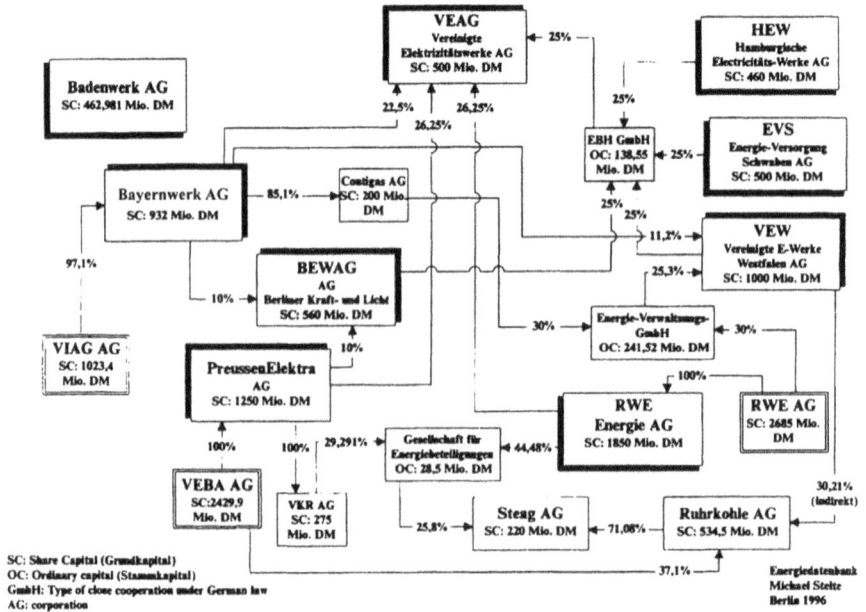

Corrected Chart Version; Original from Mez, Lutz (1997). *The German Electricity Reform Attempts: Reforming Co-optive Networks*, p. 233.

To ensure equal tariffs for rural and urban areas across Germany, the Law for the Promotion of Energy (Energiewirtschaftsgesetz, EnWG) of 1935 set up rules and sanctions for investment and price regulation. The EnWG was to prevent economic problems through harmful competition in the electric power industry, secure economical and safe electricity supply, and laid the foundation for concession treaties and exclusive monopolistic service territories. Electricity rates were defined in the Bundestarifordnung (BTO) and set up by the States.

The EnWG founded the basis for electric regulation. The energy industry was exempt from the German antitrust law (Gesetz gegen Wettbewerbsbeschränkungen, GWB). Utilities were granted permission to serve geographically closed areas through exclusive supply rights. Contracts with local city and community authorities were necessary permits to build transmission and distribution lines on the public ground. These demarcations and concession treaties protected the utility from competition within its service territory.

8.2 European Legislation for Electric Restructuring

Germany is lagging behind current restructuring efforts in the U.S. The vitalization of the European Union in regard of a common currency and free trade within the borders of Europe stimulated the restructuring debate in Germany. The discussion has begun on how to restructure the monopolistic German electric power industry. Large utilities have already started to create strategic alliances and increase their market share through mergers of large utilities (e.g. Badenwerk AG and Energieversorgung Schwaben AG in 1997) or acquisitions of smaller municipals.

In order to reduce trade barriers within the European Union, the European Commission issued guidelines (Richtlinien) to restructure the national electric power industries and open electricity markets to national and international competition. However, the guidelines of the European Union do not have the character of a binding law. They have to be transferred in national law first. European countries are now trying to integrate the European energy guidelines into national electricity laws and open up markets for competition. In January 1992, the European commission agreed to a guideline on Electricity for the European Market. Although the electric power industry tried to fight this guideline, it was not repealed. The guideline urges unbundling of electric power generation, transmission, and distribution. Fair trade and non-discriminatory competition are regarded vital prerequisites of a future liberal power market.[18]

8.3 Political Reform Attempts

In the 1970s, the German Monopoly Commission criticized the missing competition and the monopolistic service-territory structure in the electric power industry.[19] In the late 1980s, the Enquete Commission for Climate Protection reproached the incentive to ever-higher energy demand in the existing system and advocated electricity savings. Environmental concerns were also addressed by the Federal Ministry for the Environment (BMU). In March 1992, the BMU listed negative aspects of the existing EnWG such as insufficient consideration of environmental aspects in electric investment planning, incentives for increasing energy demand instead of power savings, and unsatisfactory energy efficiency in electricity generation and consumption.[20]

---

[18] Vogl, Reiner J., Manfred M. Gößl, and Gerhard M. Feldmeier (1997). *Die Elektrizitätswirtschaft in der Bundesrepublik Deutschland. Wettbewerbsstruktur im Kontext europäischer Energiepolitik*, pp. 94-108.

[19] Monopolkommission (1976). *Mehr Wettbewerb ist möglich.* Hauptgutachten 1973/75.

[20] Bundesministerium für Umwelt, Naturschutz und Reaktorsicherheit (1992). *Zur Novellierung des Energiewirtschaftsgesetzes – Defizitanalyse und Reformkonzeption aus umweltpolitischer Sicht.*

Since 1993, the BMU and the Ministry for Trade and Commerce (BMWi) released several drafts addressing electric restructuring and introducing competition to the electric power market. The drafts advocated unbundling of electricity generation; transmission and distribution; competition in the generating sector; creating pool markets; abolishing demarcation and exclusive clauses in concession treaties; as well as environmental legislation and intervention. Opponents of the drafts advocated a solution in the European context. The electric power industry stressed the importance of a long transition period toward electric competition.[21]

Large electric utilities have started mergers and acquisitions and strategic alliances with international energy companies to prepare for a liberalized European power market. German utilities develop marketing and cooperation strategies for an energy market reaching from Scandinavia (power lines to large hydropower stations are under construction) to Africa (grid-extension via Gibraltar) and Eastern Europe (nuclear power stations in Russia and Ukraine).[22]

8.4 New Energy Legislation

On April 29th, 1998, a new German Energy Law (Gesetz zur Neuregelung des Energiewirtschaftsrechts) was passed. This law adopted requirements of European legislation referring to the liberalization of the energy sector. The new German Energy Law unbundles electric generation, transmission, and distribution.[23] It abolishes protected service territories; alleviates grid access for non-utility power generators; grants priority for renewable energies in cases of transmission limitations; and guarantees feed law legislation. Renewable energy advocates criticize the 5% limit within which utilities of two vertically subsequent levels are required to purchase renewable power. Law, however, does not finally determine this limit. If the second 5% margin is reached by the subsequent utility level, the German parliament is required to issue new legislation guidelines.[24]

The new German Energy Law allows all customer categories to negotiate electric service contracts with any electric power producer. For covering transmission and distribution costs, the electric customers will have to pay an additional fee per kWh to the entities operating and owning the central grid system. Political and utility

---

[21] Mez, Lutz (1997). *The German Electricity Reform Attempts: Reforming Co-optive Networks*, pp. 244-251.

[22] Hagenmeyer, Ernst (1997). *Neue Märkte und neue Player – Strukturwandel in der deutschen Stromwirtschaft*, pp. 145-164.

[23] Informationszentrale der Elektrizitätswirtschaft e.V. (1998). *Umstrukturierung für den Wettbewerb. Bereiche Erzeugung, Transport und Verteilung werden getrennt*, p. 2.

[24] Schulz, Manfred (1998). *Marktperspektiven und ordnungspolitischer Handlungsrahmen für regenerative Energien*, pp. 3-10.

representatives are currently negotiating these transmission and distribution fees charged from industrial and residential electric customers.[25]

---

[25] Informationszentrale der Elektrizitätswirtschaft e.V. (1998). *Verbändevereinbarung über Stromdurchleitung. Durchbruch für Wettbewerb*, p. 1.

PART IV.

PHOTOVOLTAIC TECHNOLOGIES OVERVIEW

9. Photovoltaic Power Generation Components

9.1 Systematic Overview

PV solar cells are produced in standardized technical versions. Single crystalline solar cells typically produce a voltage of 0.5 Volts of direct current. The amount of electric current generated is dependent on cell efficiency, cell area and sunlight conditions. In order to achieve larger voltages and currents, groups of PV cells are combined and mounted on a metal plate, forming solar modules. Grouping modules in parallel or series to form photovoltaic panels further increases voltage and current. Photovoltaic arrays consist of a multitude of electrically interconnected solar panels. Figure 4-1. outlines the systematics of PV solar cells; modules; panels; and arrays.

Figure 4-1. PV System Components

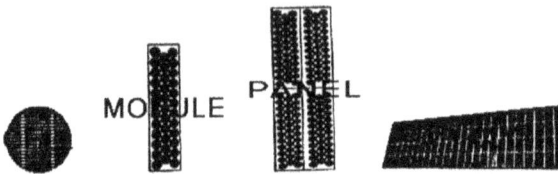

MODULE   PANEL

Gronbeck, Christopher (1998). *PV Cell Basics.*
URL http://Solstice.crest.org/renewables/re-kiosk/solar/pv/theory/physics/basics.htm.

9.2 PV Cell Types

PV cells can be manufactured from a variety of materials. Singlecrystal and polycrystal silicon is the most commonly used basic material for PV modules. These technologies have served electric power to remote and grid-connected solar homes for decades. Polycristal has slightly lower cost and lower efficiencies than singlecrystal.

Amorphous silicon is commonly used in consumer products and can be used as a thin-film technology. Module efficiencies are rather low, but provide sufficient power for watches and pocket calculators. A technical problem of amorphous silicon is its wearing-out of electrical performance; the cell structure has proven unstable over a long time period.

Concentrator technologies use optic devices (lenses) to concentrate sunlight on a small and valuable photovoltaic cell area. This helps reaching efficiencies as high as 22% today. For small applications, however, this technology is rather expensive in regard of installation and maintenance compared to conventional fixed systems without optic concentrator devices. Concentrators are therefore used for grid-connected large central PV power stations in areas of high solar radiation such as deserts.

Apart from a large array of other PV cell materials, cadmium telluride and copper indium diselenide (CIS) have the potential of significantly reducing PV module costs in mass production. These thin film technologies require but a small amount of raw material.

Table 4-1. shows an overview of PV cell technologies addressing their current global market share, module efficiencies, and manufacturing cost.[26]

Table 4-1. PV Cell Technology Overview

Technology	Global Market Share 1996	DC Module Efficiency 1995	DC Module Efficiency 2010	Manufact. Module Cost 1997 ($/kWp)	Manufact. Module Cost 2010 ($/kWp)
Singlecrystal Silicon	53.4%	15%	22%	3.30	1.10
Polycrystal Silicon	28.2%	14%	20%	3.20	1.10
Amorphous Silicon	13.2%	6-9%	14%	3.00	0.75
Crystal Silicon Concentrators	0.8%	22%	30%	4.00 (incl. tracking device)	0.75
Cadmium Telluride (CdTe)	1.8%	7-9%	14%	1.20	0.75
Copper Indium Diselenide (CIS)	N/A	7-9%	15%	1.20	0.75

[26] Maycock, Paul D. (1997). Photovoltaic Technology, Performance, Cost and Market Forecast: 1975 – 2010, pp. 1-17.

9.3 PV Module Types

PV modules are usually mounted on racks and can be installed either on the ground or on the rooftops of buildings. Racks are available for flat roofs or for tilt roofs. In most cases, installers leave some space between the roof and the PV modules for cooling purposes.

New module types try to replace the building's façade and roofing tiles. This approach helps reducing total installation costs by utilizing the benefits of PV modules as power units and façade material.

Critics have argued that PV deteriorated the design of a residential house. To enhance aesthetics, solar cells have been developed in different colors to satisfy customers and architects. However, colored cells also feature lower efficiencies than conventional cells. The most recent technical advancement is solar shingles and solar roofing tiles. They replace conventional roofing tiles and shingles and can be produced in almost any form and design. They are based on inexpensive thinfilm PV technologies. The thinfilm photovoltaic cell is integrated into the shingle made from elastic plastic material. Arrays of shingles are electrically connected by prepared wires and can be mounted easily on conventional roofs.

PV façade systems are common for large modern houses such as skyscrapers and commercial buildings. PV façade elements replace ordinary façade material and provide building protection, waterproofness, as well as electric power generation.

The following list introduces various state-of-the-art module types others than conventional standardized PV modules and discusses their assets and disadvantages.[27]

1. Solar Roofing Tiles (e.g. SolarZiegel by SolarWerk GmbH)

+ Conventional rooftop mounting is possible and can be performed by traditional roofer and tiler craftsmen.

– Usually, no additional cost reduction are possible due to using conventional roofing tiles on which the PV modules are mounted (except for special PV-integrating roofing tiles, e.g. Solar-Dach-Ziegel by NEWTEC Kunststofftechnik).

– Many wire connections reduce system reliability.

---

[27] Edinger, Raphael (1998). *Electric Restructuring and Photovoltaic Technologies - Strategies for a Deregulated Power Market*, p. 8.

2. *Solar Shingles* (e.g. SUNSLATES by ATLANTIS)

+ Using thinfilm-technologies on any flexible material; low module cost is possible.

+ Easy to integrate into building-design and retrofits.

+ Larger arrays reduce the problem of many wire connections while keeping the shingle-character of the roof.

− New technology; lifetime of solar shingles not yet determined in field applications.

3. *PV Façades* (all PV cells of any manufacturer; e.g. Pilkington Glass Company is cooperating with PV manufacturers and produces OPTISOL Solar Facades)

+ Multiple benefits of façade elements lower total system costs (e.g. daylighting, shading, and power-production all through one façade element).

+ Few wire connections through large-area arrays.

+ Conventional mounting of the glass façade is possible.

+ Use of inexpensive thinfilm technologies is possible (low production cost; low material intensity; high performance; high versatility).

− Building and façade orientation does not necessarily coincide with best solar sites.

4. *Colored PV Modules* (first introduced by BP Solar)

+ Many colors are possible (red, blue, green, brown, yellow) and can be integrated into the building's specific architecture.

− Some colors reduce the electrical efficiency of PV cells.

− Colored solar modules are more expensive due to a more complex production process.

9.4 PV System Types

9.4.1 Standalone versus Grid-Connected PV Systems

Standalone PV systems are used for remote locations where extending the central grid system would be too costly. In recent years, photovoltaic standalone systems have acquired market niches previously served by small diesel generating sets. PV systems are advantageous due to their independence from fossil fuels. Since PV modules generate electric power without any moving parts, photovoltaics have gained merits in applications where highly reliable power supply is crucial (e.g. in the space industry, for military and telecommunication applications, etc.).

Standalone PV systems have to use a back-up medium to store electricity and coordinate electric power generation and consumption. Battery systems are commonly added to the photovoltaic modules, providing electric power independently from actual current solar radiation and PV power output. However, battery systems add to the costs of the standalone power system, and the batteries' lifetime is limited (generally 5-10 years, dependent on battery maintenance, battery type, and total system design).

Grid-connected PV systems use the central electric grid as a back-up medium. The central grid provides electric power when PV system output is lower than actual power demand. On the other hand, any generated surplus energy can easily be fed back into the central grid. Grid-connected PV systems need neither battery storage capacities nor battery charging appliances and have lower costs than standalone systems.

Hybrid-systems have been developed featuring both grid-connectivity and battery-backup. These systems are designed for non-interruptible power service and are used for office and medical applications.

9.4.2 Central PV Generation versus Distributed Photovoltaics

9.4.2.1 Central PV Power Stations

Central PV power stations have been advocated for low-populated countries with high solar radiation or desert areas. The generated electric power should then be transmitted to regions with lower solar radiation but higher electric energy needs. This approach incorporates the advantages of high solar electric output in favorable solar radiation areas, but struggles with the cost for electric power transmission over long distances. Existing power lines may have to be upgraded or new lines installed to create sufficient transmission capacities for the generated electricity.

In 1985, the Southern California Edison Company connected the 6-megawatt Carissa Plains central photovoltaic plant to its power system. This project was to assess the technological aspects of photovoltaics as a central power generation option. The plant was later dismantled and the photovoltaic modules sold.[28]

---

[28] Energy Information Administration / US Department of Energy (1995). *Renewable Energy Annual 1995*, p. 111.

9.4.2.2 Grid-Support PV Substations

Photovoltaic technologies can be used for transmission and distribution support of an existing central grid system in cases of capacity congestion. In 1993, Pacific Gas and Electric (PG&E) completed the first and largest US plant to measure the benefits of grid-support photovoltaics. The Kerman substation photovoltaic plant has a capacity of 500 kWp. PG&E has evaluated the following benefits from this grid-support PV substation (selection):[29]

1. Environmental Benefits (Externalities)

Offset of 155 tons of CO_2 and 0.5 tons of NO_X each year.

2. Loss Savings

Savings of 58,500 kWh annually.

3. Substation Benefits

The PV grid-support system reduces transformer temperature and increases transformer capacity by 410 kWp. The PV grid support also extends maintenance intervals by more than 10 years and prolongs transformer lifetime.

4. Transmission Benefits

The transmission system's peak capacity is increased by 450 kWp.

We will examine the financial benefits of PV grid support in the following chapters in more detail.

9.4.2.3 Residential Grid-Connected PV Systems

The closer small power generation units are located to the customer load, the more valuable is the generated electric power. Grid-connected PV systems at residential houses minimize line losses and mitigate system congestion caused by capacity constraints in generation, transmission, and distribution. Figure 4-2. illustrates the increasing value of selected PV locations in a central grid system. Residential rooftop PV systems have a potentially higher energetic and monetary value than PV substations or central PV stations since they are located closer to the load site.

---

[29] Energy Information Administration / US Department of Energy (1997). *Renewable Energy Annual 1996*, pp. 71-74.

Figure 4-2. The Value of PV Technologies Dependent on Grid Location

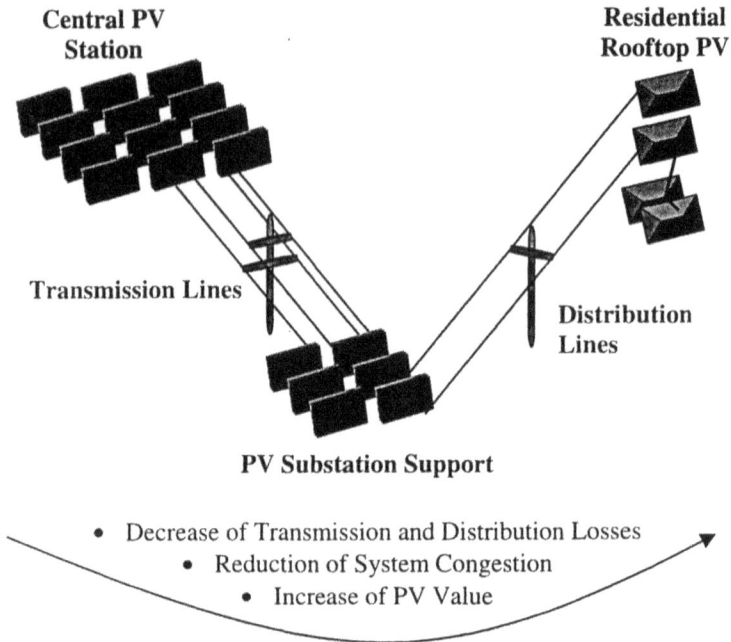

Central PV
Station

Residential
Rooftop PV

Transmission Lines

Distribution
Lines

PV Substation Support

- Decrease of Transmission and Distribution Losses
- Reduction of System Congestion
- Increase of PV Value

New England Electric Services (NEES) was one of the US electric utilities pioneering the assessment of grid-connected residential PV power systems. In 1986, NEES installed 30 small grid-connected PV systems (2 kWp each) on residential customers' rooftops. The company concluded that customer acceptance, regarding appearance, operation and output, has been excellent. The modular PV systems were installed in less than half a day by local roofers and electricians and could easily be operated by homeowners.[30]

Since the cost of PV modules was high above cost-competitiveness levels in 1986, New England Electric did not enlarge the number of grid-connected PV systems. In 1993, the California electric utility Sacramento Municipal Utility District initiated an extensive and long-term program to market grid-connected PV systems. SMUD's PV program will be outlined and discussed in the following chapters.

---

[30] Bzura, John J. (1990). *The New England Electric Photovoltaic Systems Research and Demonstration Project,* pp. 1-6.

9.4.3 Special System Types: Concentrators, Trackers, and Fixed Module Systems

Traditionally, fixed module systems were used for terrestrial photovoltaic applications. For these systems, PV modules were mounted at a fixed angle facing the sun. Energy conversion losses occurred when the moving sun did not send its radiation onto the solar cells in a vertical direction.

To increase energy conversion efficiency, concentrator and tracking devices have been developed. One-axis tracking devices follow the sun's course during daytime and increase the amount of solar radiation used for electric power generation. Two-axis trackers are even more versatile and optimize their orientation according to actual solar conditions.

Although trackers increase the electric output of a photovoltaic power system, the technical tracking device is costly and has to be maintained carefully. Dust and snow can disturb the tracker's orientation toward the sun. Trackers are usually found in utility applications. For developing countries and residential applications in industrialized nations, fixed PV systems are more common. These systems do not mechanically orient themselves to the sunrays at all times, but can be adjusted to summer and winter insolation directions.

Figure 4-3. PV Tracking and Concentrator Devices

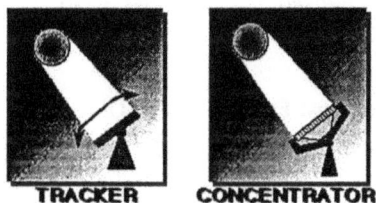

Gronbeck, Christopher (1998). *Trackers and Concentrators.*
URL http://Solstice.crest.org/renewables/re-kiosk/solar/pv/theory/track/index.shtml.

Solar equipment manufacturers have addressed the problem of costly tracking devices and expensive solar cells by developing inexpensive optic lenses that concentrate the sunlight on the PV cells' surface area. This reduces the number of solar cells required per module, thus effectively replacing cells with less expensive lenses. Figure 4-3. illustrates tracking and concentrator systems. For aesthetic reasons, the integration of concentrator devices into the architecture of residential buildings may be problematic. Therefore, concentrators are usually used for utility applications.

10. PV System Cost Comparison and Forecast

Table 4-2. and Table 4-3. present current installation costs as well as cost forecasts for various PV systems in the USA (remote systems, residential grid-connected systems, and large central power systems).[31] Remote systems prove to be far more expensive than grid-connected application for the technical reasons mentioned above.

Table 4-2. Typical PV System Cost in $/kWp (US Installations, 1997)

System Component	Remote System (1 kWp)	Residential Grid-Connected Systems (0.5-20 kWp)	Large Central Power System (200-10,000 kWp)
PV Modules (incl. Rack)	6,000	4,000	3,000
Battery	1,500	0	0
Inverter	0	750	400
Wiring, Controls	80	80	80
Labor, Profit	1,000	1,000	1,200
Total Cost ($/kWp)	8,580	5,830	4,680

Table 4-3. System Cost Forecast 2010 (US Installations, in 1997$)

System Component	Remote System (1 kWp)	Residential Grid-Connected Systems (0.5-20 kWp)	Large Central Power System (200-10,000 kWp)
PV Modules (incl. Rack)	2,000	2,000	1,000
Battery	1,000	0	0
Inverter	0	300	150
Wiring, Controls	50	50	50
Labor, Profit	800	600	300
Total Cost ($/kWp)	3,850	2,950	1,500

---

[31] Data from Maycock, Paul D. (1997). *Photovoltaic Technology, Performance, Cost and Market Forecast: 1975 – 2010*, pp. 12-13.

While large central power systems tend to have lower costs, especially regarding long-term forecasts for 2010, residential grid-connected systems located near the customer load have economic benefits that have to be taken into account. Specific information as well as a computer model on cost-benefit calculations is presented in the following chapters.

Table 4-4. indicates that the PV modules generally account for two thirds of the total system costs. This holds true for remote systems as well as for residential grid-connected and large central power systems.[32] Reducing PV module cost therefore is an important prerequisite for making PV electric generation cost-competitive with conventional power technologies.

Table 4-4. PV Module Share of Total System Cost

	Remote System (1 kWp)	Residential Grid-Connected Systems (0.5-20 kWp)	Large Central Power System (200-10,000 kWp)
1997	69,9%	68,6%	64,1%
2010	51,9%	67,8%	66,7%

Since residential grid-connected systems with fixed PV modules feature both technical and economic advantages to remote systems and represent a potentially large market with the option of mass production volumes, this study focuses on the economics of this system type.

---

[32] Calculation based on data of preceding tables.

PART V.

THE ECONOMICS OF PV POWER GENERATION IN A REGULATED AND VERTICALLY INTEGRATED ELECTRICITY MARKET

11. Costs of Photovoltaic Power Technologies

11.1 Historic Cost Decline of PV Systems

PV technologies have advanced rapidly over the last decades. Photovoltaics were first used in the space industry where high reliability and low maintenance was of high priority. In these remote locations power failure was extremely costly. With national space programs funding photovoltaic research, PV producers did not face a real incentive to decrease production cost. PV power performance, not cost, was the crucial issue in the first space applications.

This picture changed with terrestrial modules for military and telecommunication technologies. The market for solar technologies grew and PV prices decreased through learning curve effects in manufacturing experience and a more standardized production of PV modules. Between 1975 and 1995, the world price for photovoltaic modules declined from more than 80 US dollars per watt to about 4 dollars per watt (see Figure 5-1.).[33]

Figure 5-1. World Price for Photovoltaic Modules, 1973-95

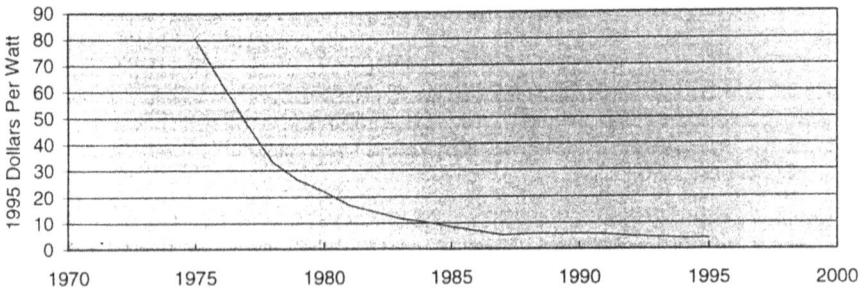

Maycock, Paul D. (1997). *World Price for Photovoltaic Modules, 1975-95.*
Worldwatch Institute Database Diskette January 1997, File Solar.wk1.

In 1997, worldwide produced module capacity reached the 100 MWp margin. 1997 was also the year when a shortage on silicon – the principal raw material for conventional photovoltaic cells – constrained the PV module production and led to increasing PV module costs.[34] The computer chip industry is the main source for

---

[33] Maycock, Paul for the Worldwatch Institute (1997), *World Price for Photovoltaic Modules, 1975-95.* Database Diskette January 1997, File Solar.wk1.

[34] Räuber, Armin (1998). *Weltweite PV-Aktivitäten – eine kritische Bewertung*, pp. 26-33.

PV cell grade silicon. The photovoltaic industry uses silicon that is of poorer quality than necessary for computer chip manufacturing. The silicon shortage was caused by the microchip industry using the raw material silicon more efficiently and thus reducing the quantity of low-grade silicon available for the PV cell industry.

In the last decades, photovoltaics have been a cost-effective option for powering residential houses located far off the grid in rural areas of the United States. Depending on the distance to the central grid system, utilities decided on a case-by-case basis if a residential photovoltaic system was a cost-effective alternative to the extension of the public grid. Sometimes, customers purchased PV systems on their own expenses when utility connection was not probable in the near future. With further cost reductions in sight, PV systems are today at the turning point of becoming cost-effective for grid-connected applications.

11.2 Break-Even System Price for Selected States

The US Department of Energy funded a study to examine break-even system prices for US States. This effort was to identify target market niches for customer-sited PV applications in the United States in order to speed the commercialization of grid-connected PV systems.

The Department of Energy's study evaluated feasible niche markets with advantageous data on the following attributes:[35]

· comparatively high retail electric rates;

· favorable tax credits and financing, leasing, and depreciation options;

· net metering options and/or rate-based incentives;

· building credits for architectural applications;

· customer's willingness to pay for clean power and innovation;

· quality of local solar resource and customer load match;

· progressive state government, regulatory, and utility support.

The study identified Hawaii to be the most promising candidate for grid-connected residential photovoltaic systems. Hawaii benefits from high solar radiation data and comparatively high electric energy rates for residential customers. Electricity rates reach up to 0.20 $/kWh and lead to break-even PV prices of 9,700 $/kWp (see Table 5-1.).

---

[35] Wenger, Howard, Christy Herig, et al. (1996). *Niche Markets for Grid-Connected Photovoltaics*, p. 1.

**Table 5-1. Electricity Rates and Break-Even PV System Price
for Residents in Hawaii**

Island	Electricity Rate (cent/kWh)	Break-Even PV Price ($/kWp)
Kauai	20.0	9,700
Hawaii – Big Island	17.0	8,400
Maui	13.8	7,000
Oahu	12.3	6,400

Figure 5-2. US States Ranked by PV System Break-Even Price

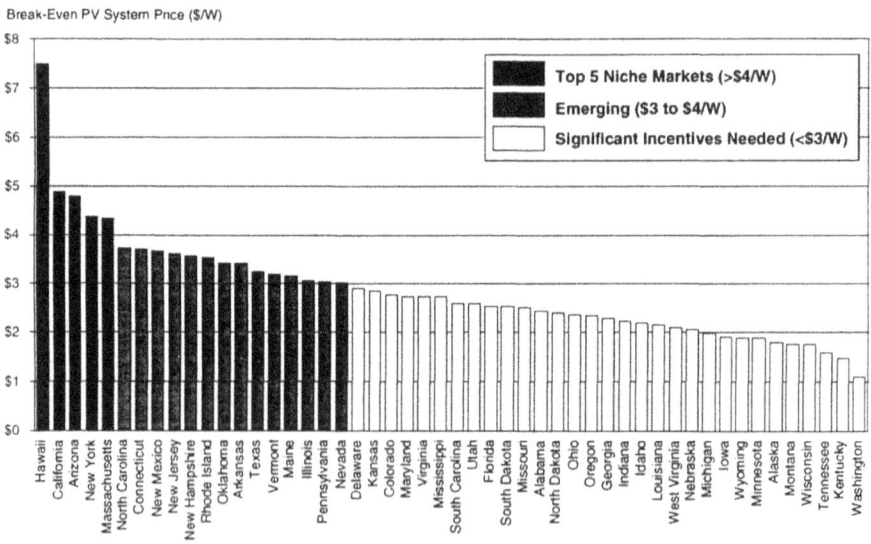

Figure 5-2. ranks US States according to their break-even PV system prices as identified by the US Department of Energy's market study. The top 5 niche markets include Hawaii, California, Arizona, New York, and Massachusetts with break-even PV prices ranging between $4,300/kWp and $7,500/kWp.

The study indicates three market tiers, with the economically most promising markets in the southwest and northeast (Figure 5-3.). The New England States benefit from relatively high residential electricity rates that makes PV power generation economically feasible in spite of the region's rather low insolation levels.

Figure 5-3. Best Break-Even Market Tiers in the Southwest and the Northeast of the USA

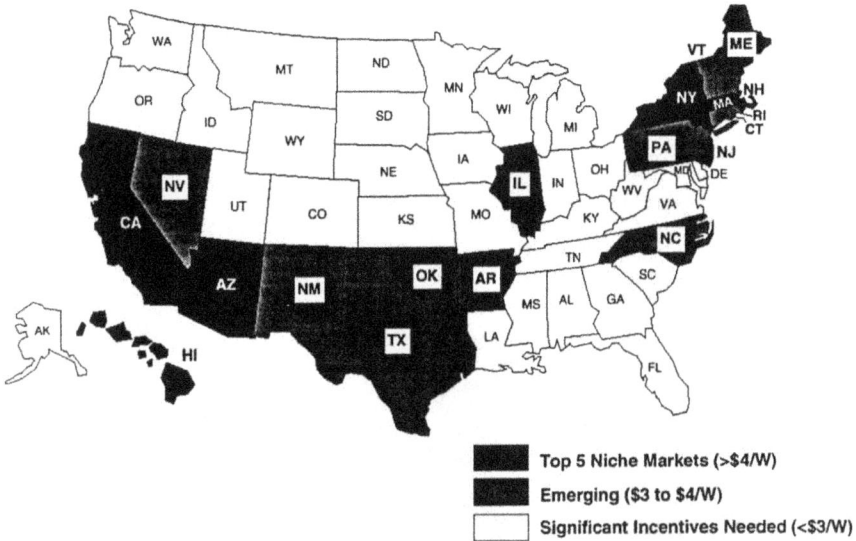

Figure 5-2. and Figure 5-3., Wenger, Howard, Christy Herig, et al. (1996). *Niche Markets for Grid-Connected Photovoltaics*, p. 3.

11.3 Projected Costs of State-of-the-Art PV Technologies

Costs for various PV cell materials as well as PV module types have been discussed in Part IV. At this point, we will examine the economics of the most promising technology in PV power systems to have an idea whether PV is feasible to reach cost-effectiveness with conventional power generation technologies.

As we have seen in Part IV, residential grid-connected PV systems have cost-advantages compared to stand-alone systems. Grid-connected systems do not require battery storage capacity since they use the public grid systems for energy back-up, if necessary. Residential grid-connected systems are expected to produce higher costs than central PV stations located in desert areas with high solar radiation. However, PV at the site of the customer is more valuable from an energetic perspective. Lower energy losses occur through on-site generation at the place of the electric consumer. Residential PV systems can be installed at the rooftops of private houses and do not struggle with finding a suitable site for a central PV power station. In the past, political decisions involving private property taxation caused unexpected financial costs to central PV stations and resulted in the bankruptcy of a US solar power enterprise.[36]

Figure 5-4. Production Cost Decline of Thinfilm-CIS-Solarcells through Mass-Production

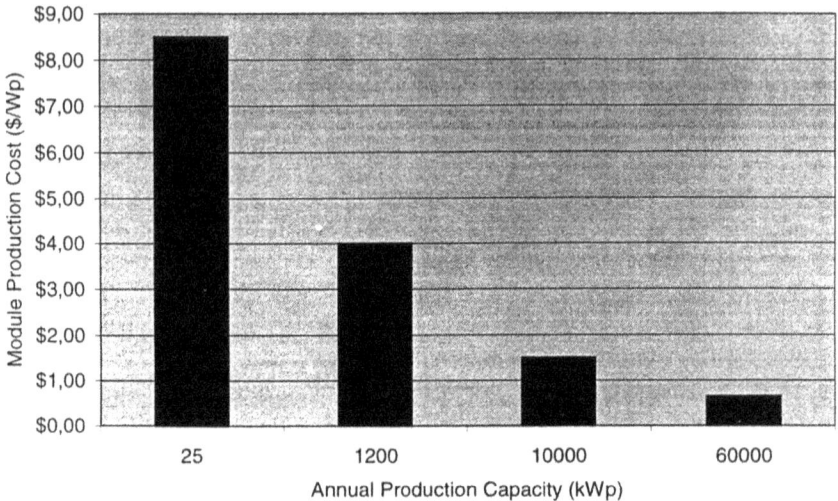

The most promising cell technology is thinfilm photovoltaic materials using only a fraction of the raw material necessary for the production of conventional solar cells. The Research Center for Solar Energy and Hydrogen (Zentrum für Sonnenenergie- und Wasserstoff-Forschung, ZSW) in Stuttgart/Germany is one of the leading institutes in research and development of copper-indium-diselenide (CIS) thinfilm PV modules and is now on the verge of commercial production. The ZSW regards a 10 MWp per annum production volume to be feasible within the next 3 to 5 years and has started to prepare mass production of its solar cells. The ZSW expects module cost as low as 1 DM (65 US cents) per Watt peak possible if annual production capacities reach 60 MWp (see Figure 5-4.).[37]

---

[36] The bankruptcy of Luz International, the first company to build a solar power station in the US, due to property taxation is described in "The Luz Experiment", Energy Information Administration / US Department of Energy, *Renewable Energy Annual 1995*, p. 102.

[37] Dimmler, Bernhard, E. Gross et al. (1997). *Progress in CIGS Large Area Thin Film PV Modules Based on Industrial Process Technology*, p. 4.

In Part IV, we have also seen that the PV modules account for the largest part of total system costs. At a share of two thirds of total system costs, the CIS thinfilm modules could cut total system costs to 1 US$/Wp.

11.4 Utility PV Cost Calculation

In the days of regulated power markets, utilities have developed cost-analysis tools to assess the value of investment options. These tools were designed for central fossil fuel power plants and do not address the specific character of modular renewable power technologies. Nevertheless, utilities utilize these investment calculations to prove the insufficient cost-effectiveness of solar energy. In this paragraph, we will examine the levelized cost calculation commonly used in the electric utility industry.

Table 5-2. Traditional Utility Investment Calculation

System Cost (US$/kWp)	$10.000,00	$7.000,00	$5.000,00	$3.000,00	$2.000,00	$1.000,00
Balance of System Cost (BOS) (1.5% of System Cost annually)	$150,00	$105,00	$75,00	$45,00	$30,00	$15,00
Energy Output (kWh/kWp)	925	925	925	925	925	925
Time Period (Years)	20	20	20	20	25	30
Discount Rate (Nominal)	0,08	0,08	0,08	0,08	0,08	0,08
Discount Rate (Real)	0,045	0,045	0,045	0,045	0,045	0,045
Cost Calculation (Nominal)						
Capital Recovery Factor (CRF)	0,101852209	0,101852209	0,101852209	0,101852209	0,093678779	0,088827433
Annual Fixed Charges (AFC)	$1.168,52	$817,97	$584,26	$350,56	$217,36	$103,83
Cost of Energy ($/kWh)	$1,26	$0,88	$0,63	$0,38	$0,23	$0,11
Cost Calculation (Real)						
Capital Recovery Factor (CRF)	0,076876144	0,076876144	0,076876144	0,076876144	0,067439028	0,061391543
Annual Fixed Charges (AFC)	$918,76	$643,13	$459,38	$275,63	$164,88	$76,39
Cost of Energy ($/kWh)	$0,99	$0,70	$0,50	$0,30	$0,18	$0,08

Remark: Annual Fixed Charge $a = K \cdot CRF = K \cdot ((1+i)^n \cdot i) / ((1+i)^n - 1)$

Edinger, Raphael (1998). *Wirtschaftliche Aspekte von Solarstrom in einem liberalen Elektrizitätsmarkt. Eine ökonomische Analyse netzgebundener Photovoltaikanlagen*, p. 2.

Table 5-2. illustrates a simple sensitivity analysis on PV system and energy costs. System costs typically range between 7,000 and 10,000 US$/kWp. This holds true for Germany and the United States according to studies of the German Fraunhofer Institute for Solar Energy Systems, Freiburg, and installations of US electric utilities assessing the economics of PV power systems. ISES evaluated lowest system cost in the German 1000-Roof-Program of 15,000 DM/kWp (=8,333 US$/kWp)[38].

Although details will be presented in the dynamic STELLA models of Part VIII, this calculation gives a first insight into the economics of PV power generation from the electric utility perspective. With system costs between $7000 and $10.000 utilities calculate energy costs from photovoltaic systems to be approximately $1/kWh. In this calculations, utilities assume balance of system costs for repairs and maintenance at annually 1.5% of total system costs; an estimated energy output of 925 kWh/kWp; a time period for system use and investment calculation of 20 years; as well as a discount rate of nominal 8% and real 4,5%, respectively.

12. Costs of Fossil Fuel Power Technologies

12.1 Costs for Electricity

Utilities use levelized cost calculations to determine the cost-effectiveness of fossil fuel investment alternatives. Conventional power generation options can be compared with this calculation method since they generally are central technologies, using large central power generators and the public grid system for transmission. They also require some kind of fossil fuel input (coal, natural gas, oil, nuclear) and have to address safety issues as well as harmful emissions or disposal of nuclear waste and other hazardous material.

The German Electric Utilities Group (Vereinigung Deutscher Elektrizitätswerke, VDEW) conducts surveys and studies on the economics of German power generation. The VDEW estimates average costs for fossil fuel power generation at 10 Pf/kWh (5.5 c/kWh) for coal fired and at 8 Pf/kWh (4.4 c/kWh) for nuclear power plants, respectively.[39] These costs are calculated for base- and intermediate load power plant operation.

Fuel costs contribute a major share to total costs per kWh. Table 5-3. shows the development of fuel costs for various fossil fuel power technologies in Germany.

---

[38] Fraunhofer Institut für Solare Energiesysteme (1995), *1000-Dächer Meß- und Ausweteprogramm, Jahresjournal 1995*, p. 4.

[39] Vereinigung Deutscher Elektrizitätswerke (1997), *Strom-Daten Februar 1997*. Kapitel 12: Brennstoff- und Erzeugungskosten.

**Table 5-3. Average Costs for Fossil Fuels
(in nominal DPf/kWh; including fuel transport)**

Year	Nuclear Fuel	Coal	Oil	Natural Gas
1985	3.0	9.5 (6.5)*	14.0	12.0
1990	2.5	9.8 (3.5)	7.2	6.8
1995	2.4	10.6 (2.9)	5.9	5.9
1997	2.2	3.1	6.2	5.9

* Cost for German coal (imported coal in brackets)

12.2 Stranded Costs

Stranded costs occur when market conditions suddenly change and existing investments are no longer cost-competitive. In a liberal market, these obsolete investments are seen as the usual entrepreneurial risk and would be depreciated in bookkeeping. Electric utilities, however, argue that they had realized these investment under regulatory advice and restrictions only. Since the government is held responsible for stranded investment, utilities call for allowance to recover stranded expenditure through higher electric rates or subsidies.

In a deregulated power market, new power providers could offer their services at lower costs than existing utilities since they are not burdened with stranded investments. Electric customers are expected to change to the supplier offering high quality service at lowest costs. Electric utilities would therefore be discriminated against through stranded investment.

In the United States, electric utilities were able to negotiate successfully the recovery of stranded investments. Regulatory commissions allowed for billions of dollars that can be recovered through higher electricity rates during a transitional time period. Stranded cost charges will be applied to usage fees for the transmission and distribution grids owned by former monopolistic electric utilities. Stranded investment can also be recovered from customers willing to change their supplier, although new suppliers may prove willing to compensate their new customers for these charges.

Estimates on the actual amount of stranded investment vary substantially. Several US research institutes have calculated stranded costs to range between US$ 70 and 200 billion (see Table 5-4.).

Table 5-4. Stranded Investment in the US Electricity Industry

	Oak Ridge National Laboratory (a)	Moody's Investor Service (a)	Resource Data International (b)
Estimated Stranded Costs (Billion US$)	72 – 102	135	202

Sources: (a) Peter Fox-Penner, *Electric Utility Restructuring.*
A Guide to the Competitive Era, p. 387.

(b) *The Electricity Journal*, April 1997, p. 4.

It is interesting to note that stranded costs occur very unevenly with the utilities assessed. Resource Data International conducted a study on 3,050 US electric utilities and estimated US$ 202 billions of stranded investment. 50% of this amount occurred with only 20 electric utilities that would face bankruptcy without the possibility to recover stranded costs.[40]

Nuclear power plants operating at higher costs than 4 cents/kWh add to stranded investments as well as long-term fuel and energy delivery contracts determining high energy costs over a long time period in a market of declining electricity rates.[41]

In Germany, stranded costs are not a crucial subject in the electric restructuring debate. This is based on the fact that large German utilities were allowed to accumulate substantial capital savings for financing nuclear waste disposal and other costs expected to occur in the future. German energy politics would not allow the recovery of stranded investments regarding these tax-exempted capital accumulations.

13. Energy Taxation and Subsidies

13.1 Subsidies for Fossil Fuels

13.1.1 Research and Development Subsidies

Direct and indirect government subsidies for fossil fuels and nuclear technologies veil the real cost of conventional power generation. Table 5-5. illustrates that more

---

[40] The Electricity Journal (April 1997). *RDI Sees 20 Utilities Holding $100 Billion in Stranded Costs*, pp. 4-5.

[41] Energy Information Administration / US Department of Energy (1997). *Annual Energy Outlook 1998. With Projections to 2020*, p. 54.

than 70 billion US tax dollars were spent for nuclear research between 1978 and 1991, and another 25 billion US dollars for fossil fuel technologies. Therefore, 85% of public subsidies were allocated for research and development (R&D) of conventional technologies.

Subsidies for R&D of renewable technologies accounted for less than 10 billion US dollars, which equals 9% of the total government research and development spending in International Energy Agency member countries.[42] In the last years, however, a growing share of R&D subsidies has been contributed to fuel cell research in the USA and Germany[43], and marketable products such as fuel cell powered cellular phones or laptop computers are on the verge of commercialization.

Table 5-5. Government Research and Development Spending in International Energy Agency Member Countries (Total, 1978-91)

Technology	Amount (billion 1991 dollars)	Share (percent)
Nuclear Fission	59.8	52
Nuclear Fusion	12.2	11
Gas Turbines	10.9	9
Other Fossil Technologies	14.4	13
Photovoltaics	2.7	2
Other Renewables	7.1	7
End-Use Efficiency	6.6	6
Fuel Cells	1	1
Total	114.7	100

13.1.2 Fuel Subsidies

The most famous German subsidy for fossil fuel power generation was the coal penny (Kohlepfennig). German electric utilities were obliged to burn German coal in order to protect the German coal industry from world price levels. The coal

---

[42] Worldwatch Institute, *Government Research and Development Spending in International Energy Agency Member Countries.* Database Diskette January 1997, File R&D.wk1

[43] The Fraunhofer Institute for Solar Energy Systems, ISES Freiburg/Germany is one of the leading research institutes involved in hydrogen fuel cell technologies.

penny was based on the "Jahrhundertvertrag" with the coal industry. This contract called for using more than 40 Mt of German coal per year.

The government subsidized German coal to mitigate economic disadvantages for the electric power industry. Table 5-3. has shown the price difference for German coal and imported international coal. With the European Union denying approval for the coal subsidy from 1993 onwards, the coal penny was abolished on January 1st, 1996. Since then, German coal is directly subsidized from the federal tax budget.[44]

13.1.3 Social Costs and Taxation of Fossil Fuel Power Generation

In a regulated environment, the government bears the risk for large accidents and nuclear waste disposal. Nuclear waste transports have to be protected by public police forces to reach their final destination. In a restructured and liberalized environment, conventional power generating technologies that bear high social risks will become more costly since they have to mitigate their risk through private insurance.

Table 5-6. Selected Externality Values as Determined by US State Public Utility Commissions (in cents/kWh)

States / Utilities	SO_2	NO_x	CO_2	N_2O	PM_{10}
California					
- Attainment Areas	0.14	0.38	0.94	-	0.03
- Southern California Edison	1.90	6.92	0.94	-	0.04
- Pacific Gas & Electric	0.36	2.01	0.94	-	0.02
Massachusetts	0.30	2.09	2.40	N/A	-
Minnesota	0 - 0.05	0.02 - 0.48	0.60 - 1.36	-	-
Nevada	0.14	1.65	2.50	N/A	0.03
New York	0.25	0.55	0.10	-	-
Oregon	-	0.44 - 1.10	1.04 - 4.16	-	-
Wisconsin	-	-	1.50	N/A	-

PM_{10} = particulate matter with diameter less than 10 microns.

---

[44] Personal Conversation with Manfred Fischedick, Wuppertal Institute for Climate, Energy, and the Environment. November 11th, 1997.

Environmental legislation such as the US Clean Air Act has remarkably reduced emission levels caused by fossil fuel power generation. To further reduce environmental damages to the air, soil, and water, regulators and governments have proposed environmental taxation to internalize social costs in the market prices of fossil fuel power generation. Therefore, financial values have been determined to quantify external costs. The State Public Utility Commissions of several US States have set up externality values as shown in Table 5-6.

13.2 Subsidies for Renewable Energies

13.2.1 German Government 1000-Roofs-Program (1990-1994)

The German Government 1000-Roofs-Program (Bund-Länder 1000-Dächer Programm) was implemented and conducted from 1990 through 1994. The program targeted the technical assessment of small-scale grid-connected residential photovoltaic systems and their integration into the public grid.

In the course of the program, 2250 grid-connected rooftop PV systems at capacities from 1 to 5 kWp were installed.[45] The investment of private electric customers ranged between 8,000 and 40,000 DM per system. The German government contributed a maximum subsidy of 27,000 DM/kWp[46].

Since the program was designed to analyze the engineering perspective of grid-integrated PV system performance, there was not a real incentive to decrease system costs. In fact, over the 5 years period, the program did not achieve a notable cost reduction. In 1991, average system costs were at 24,561 DM/kWp; in 1994, average total system costs still reached 24,021 DM/kWp.[47]

13.2.2 Investment Subsidies by German State Governments

While the PV subsidy program of the German Federal and State Governments was halted in 1994, some State governments have continued with investment subsidies for PV system installations. However, the State subsidy approach was rather unsystematic and frequently cut subsidies during the current fiscal year due to budget restrictions. For their unreliable character, these subsidies did not initiate mass production of PV technologies.

---

[45] Genennig, Bernd, and Volker U. Hoffmann (1996). *Sozialwissenschaftliche Begleituntersuchung zum Bund-Länder-1000-Dächer Photovoltaik-Programm*, p. 5.

[46] Ibid, p. 41.

[47] Ibid, p. 42.

Although the programs are frequently modified or halted, the following State programs illustrate the most typical PV system subsidies.[48] Most programs are restricted to a maximum budget; the Hamburg program subsidies, for example, have already been allocated for the 1998/99 period.

Baden-Württemberg:

Special loan program at a reduced interest rate (4% below market rate) to a maximum of 18,000 DM/kWp over a 15 years time period.

Mecklenburg-Vorpommern:

State subsidy of up to 40 % of total investment cost, maximum 18,000 DM/kWp.

Hamburg:

Cost-oriented rates in cooperation with the Hamburg Electric Utility (HEW): a) Investment subsidy of 5,000 DM/kWp and 1.30 DM/kWh (for systems 1-5 kWp) or 1.10 DM/kWh (for systems >5-10 kWp).

b) For systems without investment subsidies: 1.80 DM/kWh (for systems 1-5 kWp) or 1.60 DM/kWh (for systems >5-10 kWp), or 1.40 DM/kWh (for systems >10-50 kWp)
In each case, the subsidy is granted for a 15 years time period.

Nordrhein-Westfalen:

Grid-connected PV systems between 1-10 kWp are subsidized according to the formula **subsidy (DM) = (7.000 - 200 x peak output) x peak output**, up to a maximum of 49% of total investment cost.

Thüringen:

Investment subsidy of 8,500 DM/kWp for systems up to 2 kWp; 7,500 DM/kWp for systems over 2 kWp; up to a maximum system subsidy of 150,000 DM.

13.2.3 Subsidies by State-Owned Financial Institutes

Apart from national and State governments, some State-owned banks offer loans at lower cost if used for the purchase and installation of PV systems. The Landeskreditbank Baden-Württemberg offers special loans at an interest rate of 4%

---

[48] Deutsche Gesellschaft für Sonnenenergie (1998), *Förderung Thermischer Solaranlagen und Photovoltaik-Anlagen.* Information Brochure, pp. 1-2.

below market rate, up to 18,000 DM per kWp. The loan available for each photovoltaic system is restricted to a maximum of 100,000 DM.[49]

13.2.4 Feed Law (Stromeinspeisegesetz; since 1991)

With the Feed Law, the German government hoped to increase the share of renewable power generation. Since 1991, utilities have to buy electricity generated by small-scale solar, wind, and hydroelectric generation and are required to pay 90% of the average revenues from electricity sales, i.e. approximately 0.17 DM/kWh for small PV systems.[50]

The Feed Law proved highly successful and added large capacities of wind power generation in Northern Germany. The German utility Preussen Elektra, for example, has faced considerable competition through new wind generators at the German coastline. In consequence, German electric utilities have fought against the Feed Law.

The latest modification of the Feed Law exempts utilities from paying the 90% renewable electricity rate if the amount of current fed back into the grid from Small Power Producers (SPPs) exceeded 5% of the utility's total electricity generation volume.[51]

The subsidized electricity rates reimbursed by electric utilities through the German Feed Law are listed in Table 5-7.

Table 5-7. Subsidies for Renewable Energies Fed into the Grid (in nominal DPf/kWh)

Reimbursement Rates (1991-1998)	1991	1992	1993	1994	1995	1996	1997	1998
Wind and Solar	16.6	16.5	16.6	16.9	17.3	17.2	17.2	16.8
Biomass and Hydro-power up to 499 kWp	13.8	13.8	13.8	14.1	15.4	15.3	15.3	14.9
Hydropower 500 to 4999 kWp	12.0	11.9	12.0	12.2	12.5	12.4	12.4	12.1

[49] Internationales Wirtschaftsforum Regenerative Energien (1998). *Förderung/Energieberatung in Baden-Württemberg*. Internet http://www.uni-muenster.de/Energie/

[50] Hamburger Electricitäts-Werke AG (1995). *Ausbau der Photovoltaik in Hamburg. Technik, Wirtschaftlichkeit und Fördermodelle*, p. 54.

[51] Bischof, Ralf (1998): *Erneuerbare Energien unterm Deckel*, p. 18

13.2.5 Full Cost Rates (Kostendeckende Vergütung; since 1994)

The City of Aachen was the first German city to introduce the concept of Full Cost Rates (Kostendeckende Vergütung). Therefore, this concept is also called "Model Aachen" (Aachener Modell).[52]

Since its introduction in 1994, the full cost rates program has lead to remarkable PV capacity additions in numerous German cities. Up to date, the following cities use full cost rating or reimburse significantly higher electricity rates to small photovoltaic power producers:

Aachen, Balingen, Blomberg, Bonn, Dachau, Elmshorn, Erding, Freising, Fürstenfeldbruck, Fürth, Gießen, Gütersloh, Halstenbek, Haltern, Hamburg, Hammelburg, Heidelberg, Heidenheim, Herzogenrath, Kempten, Kiel, Ingolstadt, Lemgo, Lippstadt, Lübeck, Marburg, Moosburg, München, Nürnberg, Peißenberg, Pforzheim, Pinneberg, Raisdorf, Remscheid, Roth, Schleswig, Schorndorf, Schwabach, Schwäbisch Hall, Schweinfurt, Soest, Straubing, Traunstein, Ulm, Wedel, Würzburg, Wuppertal.[53]

Advocates of full cost rates argue that this reimbursement system is equivalent to the traditional rate-of-return regulation for the regulated power industry. The German National Utility Commission (Staatliche Strompreisaufsicht) permits utilities to recover the cost of electricity generation plus a certain profit margin. Since the generating costs of central power stations vary considerably among the technologies used with pumped storage reaching up to 1.50 DM/kWh,[54] the utilities charge a levelized price to the electric customers.

The concept of full cost rates is constructed similarly. Full cost rates cover photovoltaic system costs, interest rates for raising of capital, and a certain profit margin. Full cost rates are not based on government subsidies, but covered by electricity rates paid by the electric customer. All customers of a utility district have to participate in full cost rating and pay an insignificantly higher electricity rate. Legal studies have indicated that a maximum rate increase of 5% would be allowed by today's electricity legislation.[55]

---

[52] Solarenergie Förderverein (1997). *Das Aachener Modell*. In: Solarbrief 1/97, p. 4.

[53] Cities with full cost rates or significantly elevated rates. Source: Deutsche Gesellschaft für Sonnenenergie (1998), *Förderung Thermischer Solaranlagen und Photovoltaik-Anlagen*. Information Brochure, pp. 1-2.

[54] Solarenergie Förderverein (1997). *Das Aachener Modell*, p. 46

[55] Solarenergie Förderverein (1996). *Die Kostendeckende Vergütung*, p. 7

**Figure 5-5. PV System Installations Through Full Cost Rates
(Selected German Cities, 1994-1997)**

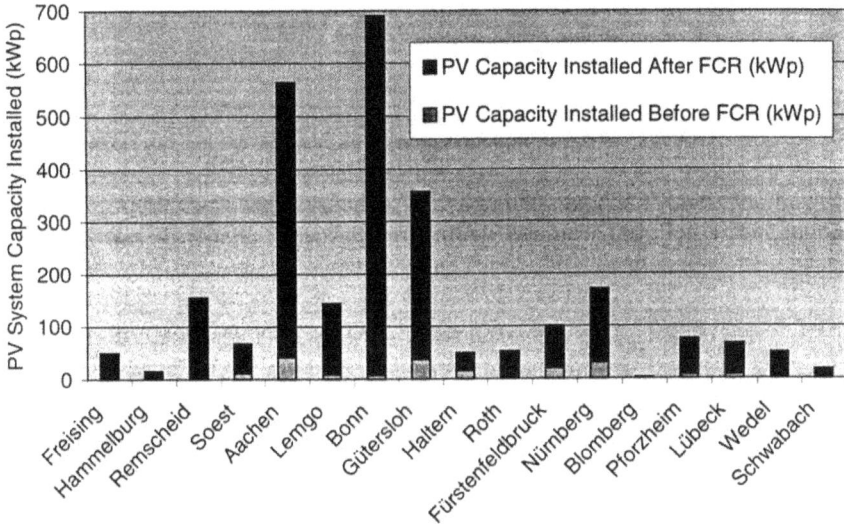

Full Cost Rates (FCR) was an attractive stimulus for private consumers to finance and operate their own photovoltaic system. While PV capacity was negligible in the cities described before FCR, the program led to up to 690 kWp installed capacity in a single city (Bonn) within a remarkably short time period (see Figure 5-5.).

13.2.6 Utility Green Pricing Programs

Apart from Federal and State programs, renewable energies have been subsidized by electric utilities themselves. 'Green Pricing Programs' were established by some utilities even before the restructuring process in order to promote renewable energies.

Most of these green pricing programs were primarily designed as pioneer programs to assess the consumers' willingness to pay more for renewable power generation.[56] However, it soon became clear that some of these programs attracted more customers than expected.

There is a wide array of green pricing program types. For *Green Tariffs*, consumers pay an incremental cents per kWh premium (e.g. Traverse City Light &

---

[56] Holt, Edward A. (1997). *Green Pricing Experience and Lessons Learned*, pp. 133-140.

Power, Michigan). The Sacramento Municipal Utility District Photovoltaic Pioneer Program, California, as well as the City of Austin Solar Explorer Program charge *fixed monthly payments* unrelated to the actual amount of energy produced or used. Customers may also nominate an amount they want to contribute to green programs and add it to their monthly bills or make an *occasional donation* (Public Service Company of Colorado and Gainesville Regional Utilities pioneered this approach in October 1993). *Rounding up* the monthly utility bill to the next dollar is a green pricing program offered by Public Services of Colorado. *Purchasing a share of capacity* in a large photovoltaic system (Detroit Edison) or even having a *utility-owned PV system* installed on the roofs of residential customers (Sacramento Municipal Utility District) are further green pricing programs used by US electric utilities.[57]

Green pricing programs do not necessarily address residential customers only, but are also marketing opportunities to commercial customers. Commercial enterprises could use their participation in green pricing programs as marketing arguments, hereby showing their commitment both to the environment and their community.

However, a regulated power market with regional monopolistic utility service territories does not offer a real choice for environmentally conscious customers to express their own preferences. When the local utility is the only electric provider, there is no opportunity for residential customers to choose between green pricing offers – they just have the option to participate in their utility's program or not. Consumers do not have the choice to subscribe to their favorite green pricing program. Electric restructuring and the introduction of retail competition have changed this restraint for the California electricity market in 1998.

The success of green pricing programs highly depends on the program's reputation of being reliable and non-discriminatory. Free-riders are a severe problem to green pricing programs. When program participants bear the full cost but cannot exclude any other resident from the benefits of an improved environment or better human health, customers could be discouraged from further participating in green pricing programs. Utilities have addressed the free-rider problem by issuing certificates or newsletters[58]. Installing the hardware such as rooftop PV systems right at the site of the customer is a promising approach to make the customer's participation in green pricing programs visible.

---

[57] Lamarre, Leslie (1997). *Utility Customers Go for the Green*, pp. 7-15.
Byrnes, Brian et al. (1996). *Green Pricing: The Bigger Picture*, pp. 18-21.

[58] The German electric utility *Dortmunder Energie und Wasser* releases and distributes the Newsletter "Airwin News" to holders of wind power certificates.

PART VI.

THE ECONOMICS OF PV POWER GENERATION IN A RESTRUCTURED AND LIBERALIZED MARKET ENVIRONMENT

14. The Threat of Electric Competition

14.1 Consumer Choice

In a regulated environment of vertically integrated power monopolists, residential customers did not have the right to choose their favorite electricity company. Each electric utility provided power to its own service territory based on concession treaties and demarcation areas.

With electric restructuring, industrial and residential customers regain their right to choose between electric service providers, as it was common in the first days of the electricity industry. Individual contracts between electric producers and power customers can be negotiated, with the electric customers paying an incremental fee on the transmission and distribution services of the grid-operating companies.

For the first time after decades of vertically integrated power markets, utilities now face competition and the threat of losing electric customers. To be successful in deregulated power markets requires utilities to adopt marketing skills and provide cost-effective electric services.

14.2 A New Market Structure

Mainly two new entities are necessary to effectively manage the emerging liberal power market. The Independent System Operator (ISO) is coordinating electricity demand and supply and grants access to the transmission network. This entity is to ensure system reliability. The Power Exchange (PX) sets the price of electric power according to supply and demand bids and works like an auction market.

Figure 6-1. Market Institutions of a Deregulated Power Market

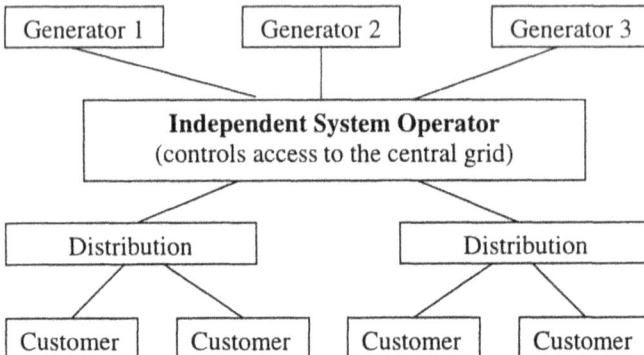

Figure 6-1. symbolizes the market coordination through the independent system operator.

The following paragraphs will introduce these two new entities in more detail for the California power market, which was opened to competition on March 31st, 1998, after a three months delay due to computer problems.

14.2.1 The California Power Exchange (PX)

The California PX is set up as a non-profit corporation. The PX sets the price of electricity (i.e. the Market Clearing Price, MCP) on an hourly basis. The market clearing price is determined by demand and supply bids from market participants who either offer or request electric power to/from the PX.[59]

The California restructuring legislation does not require all power transactions to bid through the PX. While the large investor-owned utilities Southern California Edison (SCE), Pacific Gas and Electric (PG&E), and San Diego Gas and Electric (SDG&E) have to market all their needed and offered power volumes through the PX from March 1998 through March 2002, independent power producers; municipal generators; aggregators; and electric utilities located outside of California will have the *option* of using the PX. These exempted entities can either buy and sell electricity through the PX or sell it directly to a customer. A Scheduling Coordinator supervises transactions outside the PX auction market. This coordinator is to secure transmission access and maintain system reliability through cooperation with the ISO.

The operation of the PX is controlled and regulated by the Federal Energy Regulatory Commission (FERC).

14.2.2 The California Independent System Operator (ISO)

The ISO grants or refuses access to the transmission system for all buyers and sellers of electricity. The ISO controls the dispatch of generation capacity, maintains system reliability, and balances the electric load and generation for the entire system.

While the ISO *manages* the transmission system, the electric utilities continue to *own* their transmission and distribution networks. Users of the grid system will pay a fee for electric transmission services.

---

[59] Information of the current and the subsequent paragraph (PX and ISO) is derived from Energy Information Administration EIA / US Department of Energy DOE (1998). *Industry Developments: California Successfully Starts Deregulated Electricity Market*, pp. 9-10.

In Germany, similar fees for transmission and distribution services are currently being negotiated between politicians and utility representatives. The retail customer bargains service contracts with electric power generators and pays an additional transmission and distribution charge to the communities or utilities owning relevant parts of the central grid systems.[60]

New Power Producers

In a liberalized power market, newly established power generators will compete for electric customers. Non-utility power producers were allowed to produce and consume their own electricity in vertically integrated monopolistic power markets already, but faced restrictions in selling their surplus electricity through the central grid system.

Several categories of new power providers are conceivable in a deregulated power market. First, industrial power producers will try to sell their power (mostly generated in cogeneration processes) to the market. Second, new utilities will enter the market, either national or international utilities trying to acquire new market shares. The established utilities also start to compete against each other since they lose their protected service territories.

A third category of new market participants is power marketers. These firms do not originally generate power, but purchase and sell electricity at favorable conditions. Finally, residential customers may enter the generation market offering mostly renewable power from solar or micro-hydro resources, small wind turbines, or cogeneration gas (or oil; or hydrogen) turbines.

The role of residential retail customers changes in a deregulated power market. Figure 6-2. illustrates the options for private consumers. Residential consumers may choose to stay authentic electric customers, or become small electric power producers. In the first case, consumers have the choice to change to the lowest-cost provider. The discussion on electric restructuring has excessively focussed on lowering generation costs and electricity rates. Residential customers may also find it comfortable to sign contracts-as-usual with their former utility, thus minimizing their efforts in searching for information and comparing contract offers. In this case, customers will sacrifice the potential benefits of electric competition such as lower rates or higher service quality.

Residential as well as commercial electric customers may choose to participate in green-pricing programs. These contracts have been introduced in a vertically

---

[60] Informationszentrale der Elektrizitätswirtschaft e.V. (1997). *Liberalisierung des Strommarktes: Koalition bei Energierecht einig*, p. 1.

environment already and oblige customers to pay a surcharge on their actual utility bill in order to support renewable energies.

Figure 6-2. The Role of Residential Consumers in a Deregulated Electricity Market

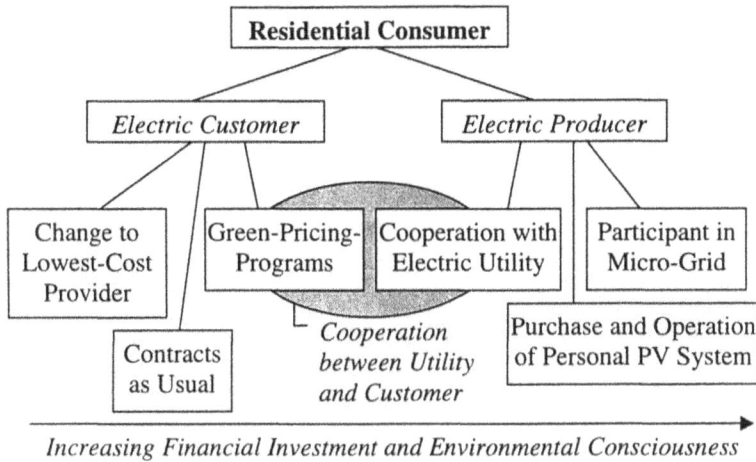

```
                    ┌──────────────────────┐
                    │ Residential Consumer │
                    └──────────────────────┘
          ┌──────────────────┐        ┌──────────────────┐
          │ Electric Customer │        │ Electric Producer │
          └──────────────────┘        └──────────────────┘

┌─────────────┐ ┌───────────────┬───────────────┐ ┌────────────────┐
│ Change to   │ │Green-Pricing- │Cooperation with│ │ Participant in │
│ Lowest-Cost │ │  Programs     │Electric Utility│ │   Micro-Grid   │
│ Provider    │ └───────────────┴───────────────┘ └────────────────┘
└─────────────┘
          ┌───────────┐   Cooperation      ┌──────────────────────┐
          │ Contracts │   between Utility   │Purchase and Operation│
          │ as Usual  │   and Customer      │ of Personal PV System│
          └───────────┘                     └──────────────────────┘
```

Increasing Financial Investment and Environmental Consciousness

Edinger, Raphael (1998). *Perspektiven der PV-Marktentwicklung in Deutschland und den USA*, p. 179.

Electric customers who want to become electric power producers may do so by purchasing their own generating systems. They may also choose to cooperate with electric utilities and combine their efforts to promote renewable energies. This kind of utility-customer-cooperation, e.g. in PV Pioneer programs, will be described below.

A group of residential customers may also cooperate and form an electrically isolated micro-grid system, generating their own electricity. These micro-grid systems may be set up and operated by residential customers or in utility-customer cooperation. The micro-grid approach will be outlined in Part VII.

Generally speaking, a deregulated power market will feature a much broader array of electric power generator types competing against each other. Their access to the central grid system and financial compensation for electric power delivered to the system will be coordinated by independent central entities such as the power exchange and the independent system operator.

15. Key Issues for Utility Investment Decision Making

15.1 Mergers and Acquisitions

The restructuring debate in the USA and in Germany has focussed on higher efficiencies as well as lower costs and prices in the electricity industry. To be cost-competitive in a deregulated environment, consulting companies have advised utilities to strive for cost-reductions through mergers and acquisitions. Larger companies are expected to have cost advantages in a liberalized environment.

US American and German electric utilities have embarked on the path of firm acquisitions and electric company mergers. The most recent example is the merger of the Southwest German utilities Badenwerk AG and Energieversorgung Schwaben AG, which now form the Energie Baden-Württemberg AG.

Mergers and acquisitions may lead to lower costs and electricity rates. It is yet unclear whether deregulation will effectively create a more competitive environment in the energy sector. Experience from the deregulation of the US telecommunication industry gives reason for concern. Since the deregulation laws were introduced, the industry has responded with an unprecedented concentration process. Seven "Baby-Bells" emerged when the US telecommunication monopolist AT&T was deregulated. Meanwhile, some of the Baby-Bells have merged again, negotiations are underway to reduce their number to four independent companies only.[61]

In a restructured energy market, mergers and acquisitions could mean a threat to electric competition, with oligopolies determining the market structure and the electricity rates.

15.2 High-Quality versus Low-Price Strategy

The concentration process of the electricity sector is caused by the industry's low-cost strategy and expectations that larger enterprises would have economic advantages in a deregulated power market. However, targeting low electric rates is not the only feasible strategy for a liberalized power market. As in any other market, both price *and* quality determine products.

At first sight, the quality of electric power does not seem to be market-relevant, because the electrons of various power generators are hardly distinguishable once they entered the central grid system. However, there are quality attributes that make electricity distinguishable and allow for marketing and branding strategies.

---

[61] Siegele, Ludwig (1998). *Es klingelt bei den Kartellwächtern. Fusionen zwischen den amerikanischen Telphongesellschaften bedrohen den freien Wettbewerb*, p. 22.

Electricity may have attributes such as low price, low outage risk (high reliability), availability[62], or environmentally friendly power generation using renewable resources such as hydropower, wind energy, or solar radiation. Contracts between retail customers and electric power producers allow charging differentiated prices for electricity of different qualities.

15.3 Customer Loyalty

In a regulated market of protected service territories, electric utilities did not face the risk of losing electric customers to other power providers (except for electric self-producing and consuming entities). This advantage ends with the introduction of competition to the electric power market. Retail customers will have the option to choose their favorite power provider. Electric utilities may lose their customers within short time periods.

Consumer choice is probably the most important challenge for electric utilities in the transformation process from vertically integrated to liberalized energy markets. Consumer orientation and customer loyalty will be crucial strategic issues in a competitive environment.

Experience from the electric restructuring process in California, however, shows that retail customers rather reluctantly make use of their new opportunity to switch to a new power provider. On March 31, 1998, 10 million California electric customers were allowed to choose their electric service provider. However, according to the Los Angeles Times, fewer than 40,000 of the 10 million eligible customers actually switched to a new ESP within the first two months of the liberalized power market. Of the large commercial and industrial customers, approximately 10 percent have switched to a new ESP.[63]

California electric customers may have learned a lesson from the deregulation of the telecommunication industry, with abundant and hard-to-compare rate information making it difficult to decide between different service contracts.

---

[62] Availability addresses the time of day / of week / of year, when electric power is available to an electric customer. Tailored rate contracts may determine the time period in which the customer agrees to restrain from consuming electric power in order to reduce systemwide peak loads, while paying lower electric rates.

[63] Energy Information Administration EIA / US Department of Energy DOE (1998). *Industry Developments: California Successfully Starts Deregulated Electricity Market*, p. 9.

15.4 Risk Assessment

Under regulated conditions, electric utilities face fewer risks since the Federal or State governments bear certain risks. If the risks have materialized, these social costs may be transferred to the retail ratepayer by increasing electric utility rates, or to the taxpayers.

In a deregulated market, risks are an integral part of competition. If not all, but most risks have to be covered by competing companies and are not as easily transferred to the social community as in a regulated environment of vertically integrated power monopolists.

Various strategic approaches try to manage the risk factors of liberalized power markets. Options and futures contracts are financial instruments to mitigate risk factors. Another possible solution is the investment in low-risk technologies. Renewable energy can help to reduce risks in a deregulated power market. Renewable power generation has economic advantages due to its modular and environmentally friendly character. New risk calculation methods try to quantify the value of reduced utility risks related to volatile fuel prices; environmental costs; construction lead-time; system modularity; power availability; initial capital costs; and investment reversibility.[64]

1. Fuel Costs

Renewables do not rely on fossil fuels. Consequently, there is no risk associated with uncertain future fuel prices. This fuel price risk can be estimated by calculating the cost for long-term fixed price fuel contracts.

2. Environmental Costs (Externalities)

While fossil fuel power plants usually consider costs for current environmental legislation and security issues when calculating levelized generating costs, additional costs for restrictive future environmental legislation and damages are typically not included. Risk-assessment tools allow estimating the probability and annual levelized cost for externalities of fossil fuel electric generation and renewable power technologies.

3. Construction Lead-Time and System Modularity

In a regulated environment, utilities usually had to satisfy projected energy demand by adding sufficient generating capacity. In a liberalized environment, future demand for a single utility is uncertain. Technologies featuring short construction lead-times have economic advantages since generating capacity can be added

---

[64] This outline is based on the work of Hoff, Thomas E. and the Pacific Energy Group (1997). *Integrating Renewable Energy Technologies in the Electric Supply Industry: A Risk Management Approach*, pp. 1-65.

incrementally and according to the actual and short-time forecasted demand. In a liberalized environment, the demand for a single utility can vary over time and become highly dynamic. Investment in large central power stations with long construction lead-times can constrain the utility's flexibility to quickly respond to current market demand. Modular renewable generation technologies help to increase the power producer's versatility and mitigate the risk of over- or under-capacities, accelerate the return on investment, and reduce the risk of project-delays.

Modular power generators can be installed in increments and at locations where capacity constraints are the most urgent or the value of electric power is the highest (i.e. near the customer load site). Total initial capital cost is lower for modular systems, therefore fewer capital resources are tied up during shorter construction lead-times. This advantage enhances the electric utility's financial liquidity.

4. Investment Reversibility

Large central power stations are expected to operate for long time periods and tie up large capital volumes. If the plant becomes economically obsolete before recovering investment costs, the reversibility of the investment is a crucial financial factor. Large power stations are usually specifically designed for their operation environment and are rather difficult to sell or move to other sites. Modular plants can be easier relocated to sites of higher generating value or be sold for other applications. In the case of photovoltaic power generation, the modular photovoltaic modules can be disassembled and moved to areas of higher solar radiation or higher demand for solar energy, or may even be sold to commercial or residential customers.

5. Power Availability

Modular power generation has economic advantages over large central stations in regard of availability issues. Modular power plants can start to produce power after each segment's completion, while central power stations have to be completed as a whole before producing and selling electric power.

Furthermore, equipment failures lead to outages of the complete central power plant, while outages of a single modular power unit only reduce a part of the distributed system's generating capacity.

The categories mentioned above will be monetarized from a selected utility's perspective in Part VII and used for a model of distributed power generation in Part VIII.

16. Strategies for Photovoltaic Power Generation

16.1 Customer Support for Renewable Energies

In industrialized countries, public support for a clean environment is growing. Opinion polls in the US conducted since the 1970s have shown increasing customer support for energy conservation, energy efficiency, and renewable resources augmenting from less than 40 percent to more than 70 percent of the participants in recent surveys.[65]

There is a considerable number of customers who are even willing to pay higher prices for clean energies, although the restructuring debate has often concentrated on how to decrease generation costs and electric prices. Consumers seem to value the quality component of clean renewable energies. Utility green pricing programs try to address this valuable market segment.

Figure 6-3. RWE Market Survey "Willingness to Pay a Premium for Renewable Power Generation" (in DM per Month)

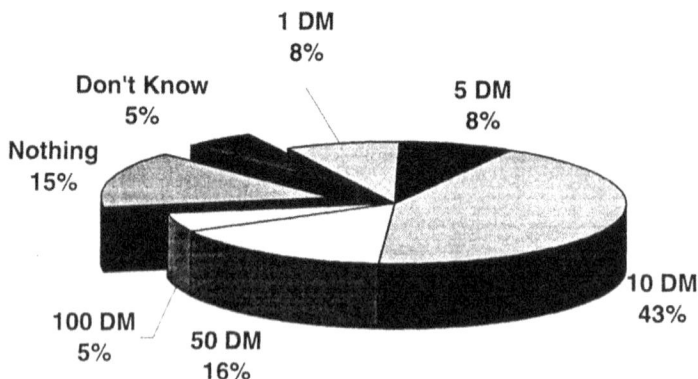

German electric utilities have conducted market surveys to assess the potential of consumers interested in renewable energies. The largest German electric utility RWE has asked its customers whether they were willing to pay extra for clean power generation, and if so, how much per month. The survey results are illustrated in Figure 6-3. (in German Marks; 1 US$ approx. 1.80 DM). With 20% of the

---

[65] Utility PhotoVoltaic Group (1996), *Multi-Utility Market Survey on "PV Friendly" Pricing Reveals Strong, Positive Reaction to PV*, p. 1.

participants voting "Don't know" or "Nothing", a considerable 80% of the surveyed people were in favor of paying an extra premium for renewable power generation per month. Since the survey was not conducted with the threat of having the participants actually pay the premium, these results have to be regarded carefully. The RWE green pricing program was not able to acquire more than 0.5% of total RWE residential customers.[66]

16.2 Marketing and Branding: Selling Clean Energy

Marketing green power is a more holistic approach compared to utility green pricing programs. Green pricing programs target consumers willing to pay a premium for environmentally friendly power generation. The utilities argue that renewable power generation is not and will not be cost-competitive with conventional power generation technologies in the short- and mid-term future. Green pricing programs can be initiated by utilities generating only a small share of their total power capacity from renewable resources.

Marketing green power is a strategy for utilities that regard renewable power generation as their core business. The utility name can be used as a brand name. With this approach, the electric utility does not principally aim at charging higher than average market prices for its electric power. We will see below that strategic marketing efforts of green energies may finally lead to cost-competitive renewable energy prices.

Treating green power as a brandable product has initiated German non-profit organizations to set up general certification guidelines and eco-power labels. This approach is meant to introduce general quality-criteria to enhance customer transparency for green power products. Currently, two categories of green labels are being discussed: the golden eco-label for electric power generation from 100% renewable resources, and the silver eco-label for electric power generation using an additional share of cogeneration.[67] The labels and certificates are set up by non-profit non-utility organizations such as EUROSOLAR and the TÜV Rheinland[68] and will be available to any utility and non-utility power producer interested in marketing green power.

---

[66] Beyer, Ulrich (1998). *Konzeption und Kundenerfahrung mit dem Umwelttarif der RWE Energie AG*, pp. 23-29.

[67] Pontenagel, Irm (1998). *Präsentation von Label-Ideen für exclusive Öko-Stromanbieter und Öko-Stromkunden.* Presentation at the EUROSOLAR Conference „Der Öko-Strommarkt", Haus der Wirtschaft, Stuttgart, April 25th, 1998.

[68] Wiesner, Wolfgang (1998). *Zertifizierungsgrundlage für Anbieter und Kunden des Öko-Strommarktes.* Presentation at the EUROSOLAR Conference „Der Öko-Strommarkt", Haus der Wirtschaft, Stuttgart, April 25th, 1998.

16.3 Utility-Customer-Manufacturer-Cooperation

16.3.1 System Cost Reductions

The most important barrier to photovoltaic power generation is the high up-front investment costs for PV modules. Technological progress has remarkably reduced PV system costs, as discussed in Part V. However, at today's level of total system costs, photovoltaic power generation is generally regarded to be far from being cost-competitive with conventional electric generation technologies.

Various PV cell technologies are assessed on their contribution to production cost reductions. Considerable cost savings are expected from mass production. PV manufacturers have not yet installed large enough production sites.

Photovoltaic power systems for residential customers are a promising market that could encourage PV manufacturers to invest in large automated assembly plants. Electric utilities interested in promoting renewable power generation could function as PV technology marketers, purchasing large and reliable quantities of PV modules for residential grid-connected PV installations. In a liberalized environment, the electric utilities could themselves profit from this strategy, as we shall see below.

16.3.2 Long-Term Customer Contracts

In a deregulated power market, electric customers have the right to choose their favorite power supplier and switch between competing utilities. As experiences from the deregulated US telecommunication market shows, companies may lose and gain customers within very short time periods, which sets the electric utility's financial revenues at risk. Protected service territories no longer exist to secure the utility's customer basis.

Long-term customer contracts are highly valuable assets in a liberalized electric power market. Utilities that are able to acquire customer contracts over a 12-months or even a several years time period will have a vital competitive advantage in this highly dynamic power market.

16.4 An Empiric Example of a Utility Photovoltaic Market Strategy

16.4.1 A Municipal Utility's Case Study: From Nuclear Power Generation to Pioneering Solar Energy

The Sacramento Municipal Utility District (SMUD) is a remarkable example of shifting from conventional centralized generation technologies to advanced solutions of energy services. Today, SMUD is the leading US municipal utility in regard of residential customer orientation, demand side management efforts, and photovoltaic marketing strategies.

SMUD is the fifth largest US municipal utility in terms of energy sales, serving approximately 475,000 customers in a 900 square miles service territory in and around the City of Sacramento, the capital of California. A seven-member Board of Directors elected by the public for four-year terms governs the municipal utility. The board is responsible for the District policy and setting electric rates.[69]

In 1985, a serious accident occurred at SMUD's nuclear reactor "Rancho Seco" on the day after Christmas. After a crimped wire in an electrical switching box had shut down the central computer, the operators lost control of the 913-megawatt reactor that provided most of SMUD's electric power. This failure led to an unexpected sequence of emergency situations; an accident similar to the Ukrainian Chernobyl could have very likely occurred.

An analysis of the damage, necessary repair, and newly required security features for Rancho Seco determined that that a repair was not economical. SMUD searched for an interested investor taking over the damaged nuclear power plant, but was not successful. Finally, SMUD decided to decommission Rancho Seco.

After the accident, SMUD was in need of purchasing power due to sudden under-capacities. Extensive market studies and consumer surveys indicated that SMUD's customers favored demand side management programs and investment in renewable energies over conventional power generation technologies.

SMUD initiated an unprecedented demand side management campaign. A program to plant new trees for shading residential houses helped to cut peak summer loads by reducing active air conditioning. Installation of high-efficiency air conditioners, solar hot water heaters, and compact fluorescent lamps as well as incentives for efficient refrigerators were additional approaches to use energy more efficiently. By the end of 1995, these programs had resulted in reducing peak demand by 372 megawatts, which equals about 40% of the capacity of the Rancho Seco nuclear power plant. SMUD's demand side management (DSM) programs proved to be very popular with its customers, with half of SMUD's customers having participated in at least one DSM program until 1996.[70]

Apart from its demand side management program, SMUD started to assess renewable generation options for their ability to contribute cost-effectively to its generating portfolio. In 1993, SMUD launched its 'PV Pioneer Program' for residential customers. SMUD installed grid-connected rooftop PV systems for a

---

[69] The Results Center, *Sacramento Municipal Utility District Solar Photovoltaic Program, Profile #111*, pp. 2-3.

[70] Smeloff, Ed, and Peter Asmus (1997). *Reinventing Electric Utilities. Competition, Citizen Action, and Clean Power*, pp. 53-56.

monthly green fee of an additional 4 US$, which is approximately a 15% premium to the customers' average utility bill. SMUD purchases, installs, owns, and operates the residential rooftop PV systems, sized at about 4 kWp each. The PV Pioneer Program proved more successful than expected. Each year, about 500 to 1000 customers volunteer for the approximately 100 PV Pioneer systems available annually. Until 1997, SMUD had installed about 1.9 MWp photovoltaic capacity at 436 residential and commercial rooftop systems.[71]

16.4.2 A Strategy for Photovoltaic Power Generation

The SMUD engagement in solar energy profoundly differs from other utility's solar programs. SMUD's photovoltaic strategy is not designed to examine technological aspects of grid-connected PV applications only. The PV Pioneer program does not aim at installing photovoltaic power systems on a trial-and-error basis and planning to abandon the program if other technologies would promise a higher return-on-investment. SMUD developed a strategic marketing concept to create long-term electric customer relationships and actively bring down photovoltaic system costs.

The municipal utility's Sustained Orderly Development (SOD) concept focuses on two integral aspects of marketing solar electric power. First, the strategy addresses residential customers interested in renewable power generation. The PV pioneer program creates long-term customer relationships with the utility installing photovoltaic systems on residential houses and private residents signing electric service contracts for 20 years. The second aspect of the Sustained Orderly Development strategy refers to reliable long-term contracts with the photovoltaic industry. Larger shipments of photovoltaic systems are commonly regarded necessary to accelerate commercialization through mass production of PV systems and to make manufacturers independent from any form of subsidies. In course of SMUD's SOD strategy, the electric utility managed to have a local PV equipment manufacturing plant open in its service territories, with advantages to both the utility and the PV manufacturer. While the guaranteed PV module orders allow the system manufacturer to invest in larger production and assembly sites, the electric utility profits from designated system cost reductions set up in the order and purchase agreements.

The long-term contracts with the PV industry have effectively reduced SMUD's solar system costs for grid-connected rooftop applications. SMUD regularly announces the volume of its PV system procurement to the PV manufacturing industry and receives offers in form of competing bids. SMUD then selects and chooses the most suitable bid for its photovoltaic system projects. In the following

---

[71] Osborn, Donald E. (1997). *Commercialization of Utility PV Distributed Power Systems*, pp. 1-7.

figures, data points are based on historic system costs and actual bids received by SMUD for the 5-year, 10 MWp purchase between 1998 and 2002. SMUD expects this multi-year effort to lead to a new PV manufacturing facility in Sacramento.[72]

Figure 6-4. PV System Cost Development and SMUD's SOD Strategy

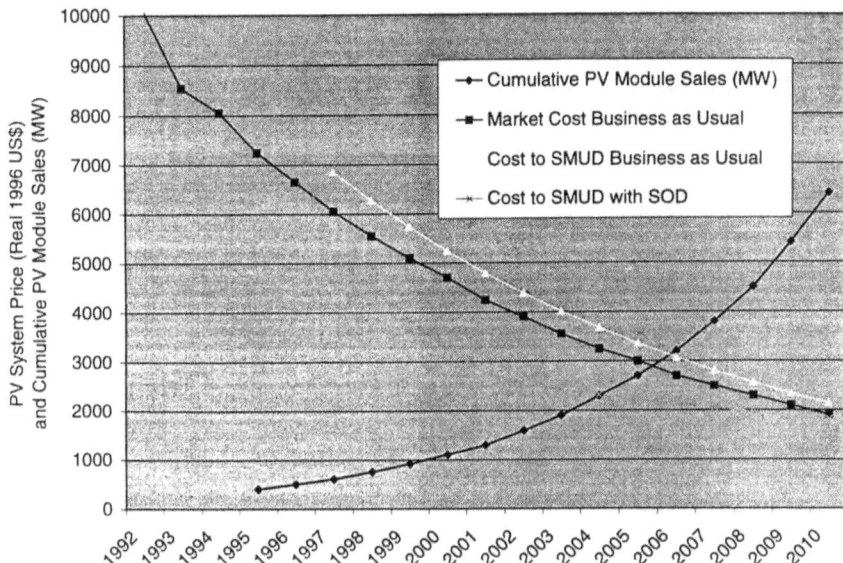

Figure 6-4. and Table 6-1. outline the cost-effects through the SOD strategy and demonstrate SMUD's long-term commitment and strategic PV market orientation. The yellow curve shows PV system costs to SMUD being higher than market costs in a business-as-usual scenario. In 1993, SMUD started its SOD strategy and PV Pioneer program and in the beginning accepted higher than market costs to the utility (see Table 6-1.: market costs of 8,550 $/kWp compared to SMUD's SOD costs of 9,800 $/kWp in 1993). SMUD's PV system costs did not break-even with market costs until 1997. However, from 1997 onwards, the municipal utility has a competitive advantage through its PV system costs being lower than market costs.

[72] Osborn, Donald E. (1997). *Commercialization of Utility PV Distributed Power Systems*, pp. 153-159.

Table 6-1. PV System Cost Development and SMUD's SOD Strategy
(table values in real 1996 US$/kWac)

	Cumulated PV Module Production (in MWp)	Market Costs Business as Usual	Cost to SMUD Business as Usual	Cost to SMUD with SOD
1992		10,100		
1993		8,550		9,800
1994		8,050		9,260
1995	400	7,250		8,360
1996	500	6,650		7,580
1997	610	6,050	6,850	5,100
1998	750	5,550	6,270	4,530
1999	910	5,100	5,730	3,850
2000	1,100	4,700	5,240	3,250
2001	1,300	4,250	4,790	2,770
2002	1,600	3,900	4,380	2,540
2003	1,900	3,550	4,010	2,420
2004	2,300	3,250	3,670	2,320
2005	2,700	3,000	3,350	2,220
2006	3,200	2,700	3,070	2,140
2007	3,800	2,500	2,800	2,050
2008	4,500	2,300	2,560	1,970
2009	5,400	2,100	2,340	1,880
2010	6,400	1,900	2,140	1,810

SMUD expects its grid-connected photovoltaic system costs to fall below 3,000 $/kWp in 2001, which would make PV a cost-competitive option for the California electricity retail market where prices reach up to 10 cents per kWh today. In 2010, SMUD calculates its PV systems to be cost-effective even without subsidies (see Figure 6-5.).[73]

---

[73] Wenger, Howard, Thomas E. Hoff, and Donald E. Osborn (1997). *A Case Study of Utility PV Economics*, pp. 171-176.

Figure 6-5. SMUD's Sustained Orderly Development Strategy (SOD) to Actively Bring Down PV System Costs

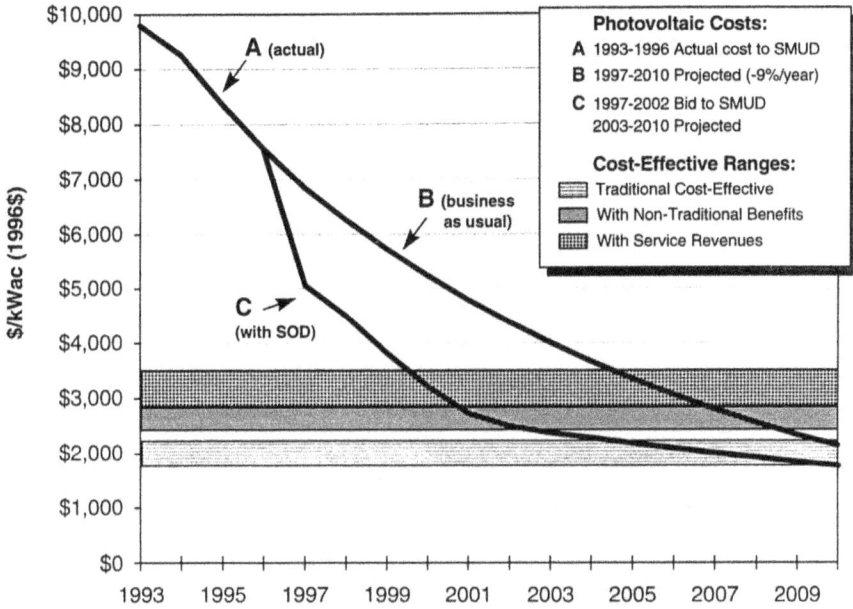

Wenger, Howard J., Thomas E. Hoff, and Donald E. Osborn (1997).
A Case Study of Utility PV Economics, p. 175.

16.4.3 Solar Energy Strategies and Electric Restructuring

Other utilities are just about to launch similar projects. The City of Austin set up its Solar Explorer project for residential customers.[74] The electric utilities benefit from these residential rooftop PV projects in various ways. They acquire experience in new modular power generation technologies and in using renewable resources. With their installations of grid-connected residential rooftop PV systems, they show their commitment to social and environmental concerns, which can add to the utility's corporate identity strategy in a restructured and liberalized market.

Regarding the concept of distributed power generation, there are also economic benefits for the utility investing in small-scale photovoltaic systems in order to defer grid or generating capacity investments. With consumers offering their own rooftop for free use, the electric utility can avoid long and expensive search for new

---

[74] Libby, Leslie (1997). *Austin's Solar Explorer Program*, pp. 161-163.

power plant locations and thus avoid regulatory permission procedures, which in turn accelerates the lead-time of installing new generating capacity.

Utilities signing photovoltaic service contracts with residential customers are building long-term consumer relationships and enhancing customer loyalty, which is a valuable competitive asset in a dynamic liberalized electricity market. Installing utility-owned photovoltaic systems on residential rooftops enhances utility-customer relationships and creates a stable customer basis to the electric service company.

PART VII.

DISTRIBUTED POWER GENERATION:
CREATING A DECENTRALIZED ELECTRICITY SYSTEM

17. The Concept of Distributed Generation

17.1 The Decline of Economies of Scale in Electric Power Generation

Experts are not yet clear on whether technological advancement has initiated electric deregulation or vice versa. Analysts believe that electric restructuring has been driven by the remarkable divergence between the price of electricity and the marginal cost of new capacity.[75] Technology and economics of today's state-of-the-art electric power generation technologies differ considerably from conventional power systems.

Figure 7-1. Optimal Plant Size (per MWp Cost Curves 1930 – 2000)

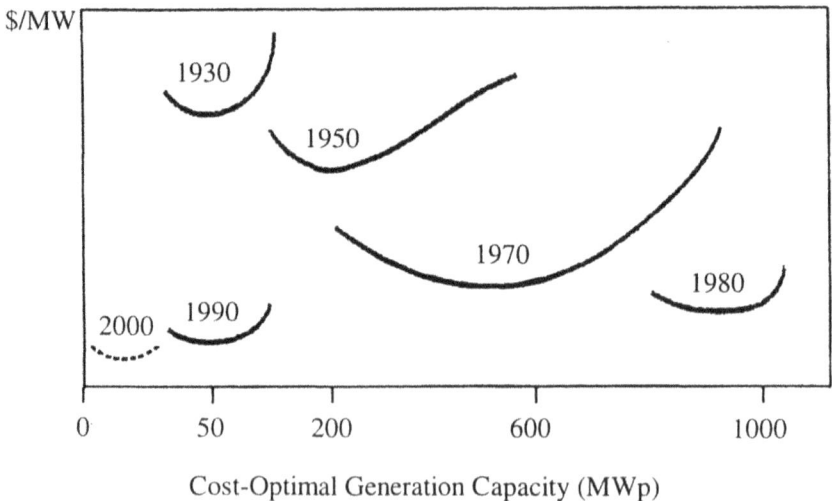

Cost-Optimal Generation Capacity (MWp)

Elaborated Chart. Original from Bayless, Charles E. (1994). *Less is More: Why Gas Turbines Will Transform Electric Utilities*, p. 24.

Until the 1980s, large power plants up to 1000-megawatt capacity continuously decreased the average cost of electricity generation through economies of scale. The emerging of high-efficiency gas turbines reversed this trend (see Figure 7-1.). Modern combined-cycle gas turbines reach their maximum performance in power

---

[75] Letendre, Steven, John Byrne, and Young-Doo Wang (1996). The Distributed Utility Concept: Toward a Sustainable Electric Utility Sector, p. 7-2.

plants as small as 400 megawatts, and aero-derivative gas turbines are efficient at 10 megawatts.[76]

Gas turbines reverse the trend to ever-larger central power stations. They can be installed incrementally and at locations near the customer's load. Due to their small-scale and modular character, gas turbines feature comparatively low construction costs per generation unit and short lead times. With projections of declining gas prices in the near future, gas turbines are regarded a valuable asset and investment option to achieve cost-reductions for electric power generation in a deregulated power market.[77]

Decreasing economies of scale in central power generation will lead to a more decentralized electricity system based on a higher number of smaller generation units distributed across the utility grid.

17.2 Defining Distributed Power Generation

Distributed power generation (or short: Distributed Generation, DG) refers to small-scale electric generation units that are located at decentral locations across the central grid system. Distributed generation is therefore located near the customer load and minimizes grid losses. DG may also be set up as separate systems independent from the central grid. Remote residences producing and consuming their own power account for this category of DG, as well as a group of houses forming their own grid system, which will be discussed below as micro-grid systems.

The Interstate Natural Gas Association of America Foundation (INGAA) defines Distributed Generation as "any small scale power generation technology that provides electric power at a site closer to customers than central station generation, and is usually interconnected to the transmission or distribution system". In the future this definition may change from "is usually interconnected" to "which may or may not be interconnected".[78]

17.3 Distributed Generation with Renewable Energies

Next to the gas turbine technologies mentioned above, a distributed energy system may contain a variety of small-scale renewable energy sources such as solar

---

[76] Balzhiser, Richard E. (1996). *Technology – It's Only Begun to Make a Difference*, pp. 32-55.

[77] Energy Information Administration / US Department of Energy (1997). *Annual Energy Outlook 1998. With Projections to 2020*, p. 63.

[78] Howard, Milton R. (1996), *Advancing Electric Competition by Providing Electric Power Choice*, pp. 1-2.

energy, microhydro and wind power, as well hydrogen fuel cell technologies. Distributed generation and storage as well as demand-side management efforts amplify central electricity generation.[79] The multitude of small power generators may also add to the system's reliability, with outages less likely to occur than in a system with one major generation unit.

Large utility-scale and small residential wind turbines can be installed near villages or at the electric consumer's site. Residential PV systems can be mounted on rooftops of private houses. When floating water is available, medium-scale or micro-hydro turbines can add to the distributed power system providing cheap electric energy.

17.4 Utility Benefits from Distributed Electricity Generation

17.4.1 Economic Assets through Distributed Power Systems

In the past, electric utilities normally responded to demand increases with adding new generation capacity, as shown in Figure 7-2. In consequence, the vertically integrated utility had to upgrade its transmission and distribution system, too.

Figure 7-2. Traditional Utility Response to Demand Increases: Building a New Generation Facility and Transmission Network

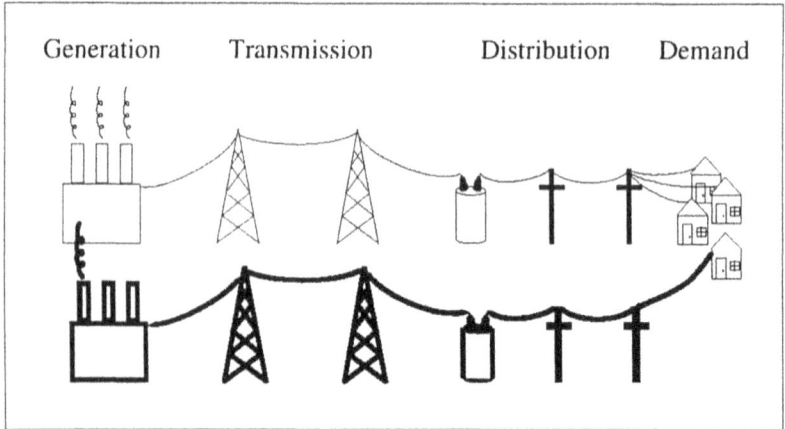

Hoff, Thomas E., Howard J. Wenger, and Brian K. Farmer (1996). *Distributed Generation: An Alternative to Electric Utility Investments in System Capacity*, p. 137.

---

[79] Letendre, Steven, John Byrne, and Young-Doo Wang (1996). *The Distributed Utility Concept: Toward a Sustainable Electric Utility Sector*, pp. 7-1 to 7-8.

Distributed electricity generation provides electric power at sites where capacity constraints in generation, transmission, or distribution are the most urgent. The most valuable location for electric power generation is near the customer's load. Distributed generation allows for smaller central power capacities. DG thus achieves financial savings by deferring the construction of new facilities; transmission lines; substations; feeders; and by reducing line losses with electricity production near the customer load (see Figure 7-3.).[80]

**Figure 7-3. The Value of Distributed Generation to the Utility System:
1. Deferring Capacity Investment**

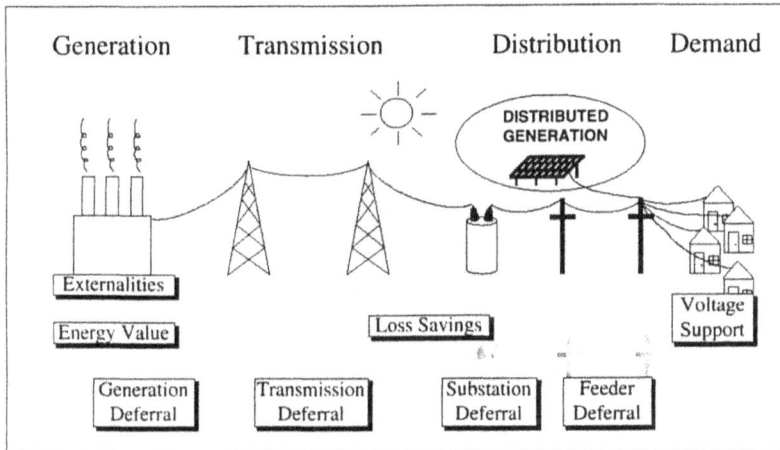

Hoff, Thomas E., Howard J. Wenger, and Brian K. Farmer (1996). *Distributed Generation: An Alternative to Electric Utility Investments in System Capacity*, p. 147.

Investment costs for large central power stations and the electric grid system could be easily recovered in an environment where state regulators allowed for capacity additions and agreed to electricity rates that would recover investment costs.

In a deregulated electricity market, electric rates will be determined by average market prices set by demand and supply. Utilities may find it uneconomic or at high risk to invest in large central power generation technologies when the utility is yet unclear about its future customer basis and energy demand. Investing in large-scale generation technologies constrains the utility's flexibility and determines

---

[80] Hoff, Thomas E., Howard J. Wenger, and Brian K. Farmer (1996). *Distributed Generation: An Alternative to Electric Utility Investments in System Capacity*, pp. 137-147.

generation costs for a long time period. Distributed power technologies are advantageous in a liberalized power market due to their modular character since they can be installed incrementally according to the actual demand growth. The electric utility can avoid uneconomic over-capacities, which would have occurred by installing a new central power plant that had to satisfy anticipated future demand increases (see Figure 7-4.).

**Figure 7-4. The Value of Distributed Generation to the Utility System:
2. Reducing Excess Capacity**

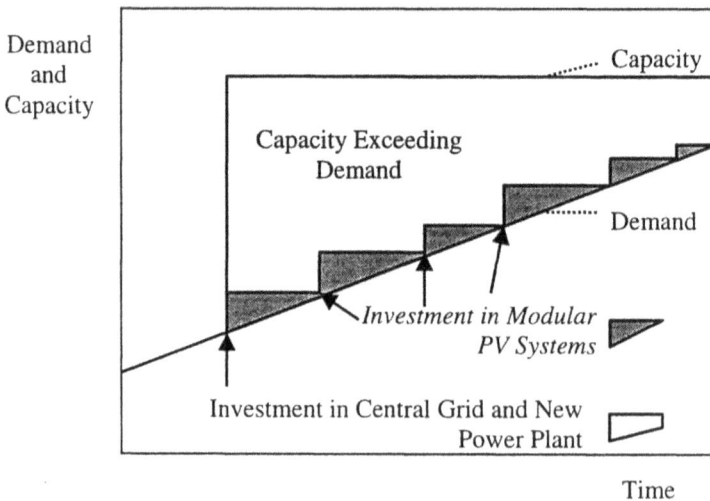

An electric utility may also face the adverse case of under-capacities due to underestimated future energy demand. Building large central power capacities would therefore delay the utility's return on investment, since power sales would not be possible until the construction is completed. Distributed generation technologies can accelerate the utility's return on investment as shown in Figure 7-5. This figure nicely demonstrates the two cases of over- and under-capacities due to erroneous expectations in electricity demand. The right side of the chart shows the case of overestimated demand growth. The electric utility has installed a central power station at a size larger than necessary to satisfy future demand increases. Costs for the central power station occur before the plant is operational, with planning and construction times (lead-times) for central power stations reaching up to a decade. Until the actual demand reaches the plant's capacity, the utility will not be able to reach maximal revenues. In contrast, the costs of small generating units (symbolized by small dark-gray triangles) occur incrementally. These systems are capable of recovering their maximum revenues sooner than the large power plant.

The left side of the chart reflects the scenario when utilities have underestimated future demand growth and have not started to add new capacity in time. The utility now faces two options: either building a large central power plant that would not deliver power until its final completion, or adding small modular generating units incrementally. The latter case would reduce total up-front investment cost until the first possible power sales. With lead-times of several days (for photovoltaic systems) or weeks (for microhydro systems), the utility would be able to start power sales earlier and sooner receive revenues.

**Figure 7-5. The Value of Distributed Generation to the Utility System:
3. Accelerating Return on Investment**

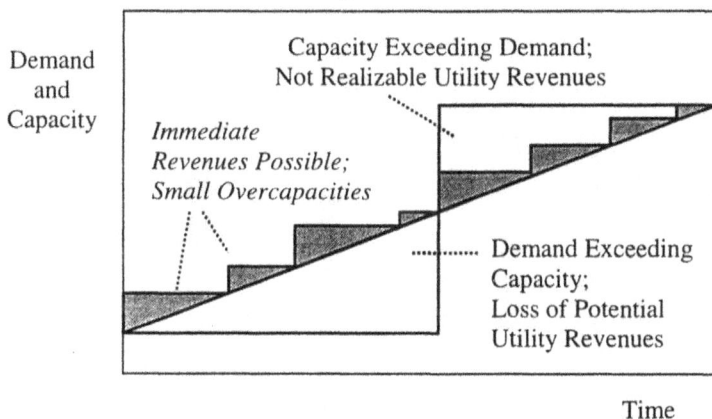

17.4.2 **Valuing Distributed PV Power Generation**

For the reasons presented above, distributed generation can have economic advantages over central power technologies. Advocates for distributed systems and renewable energies call for using cost-benefit analyses instead of conventional levelized cost calculations. Utility-specific cost-benefit calculations value the assets of small-scale distributed technologies in a central grid system.

The chapter on dynamic modeling (Part VIII) will present the cost-benefit categories and monetary values as assessed by the Sacramento Municipal Utility District (SMUD) for its grid-connected residential photovoltaic systems. At this point, we present a more general method to assess utility-specific benefit-values of distributed technologies.

The Pacific Energy Group developed a computer software package called QuickScreen under contract to the US Department of Energy. QuickScreen evaluates the economics of specific distributed PV applications from an electric utility's perspective.

QuickScreen captures eight categories of distributed PV system benefits, such as externalities; savings in distribution, sub-transmission, and bulk-transmission; energy; generation capacity; and "other". With the "other" category, a utility can add its specific benefits from PV investment (e.g. local tax advantages). Electric loss savings are accounted for in each of the relevant categories.[81]

Figure 7-6. Utility Photovoltaic Benefit Analysis: Detailed versus Quick-Screen Calculation

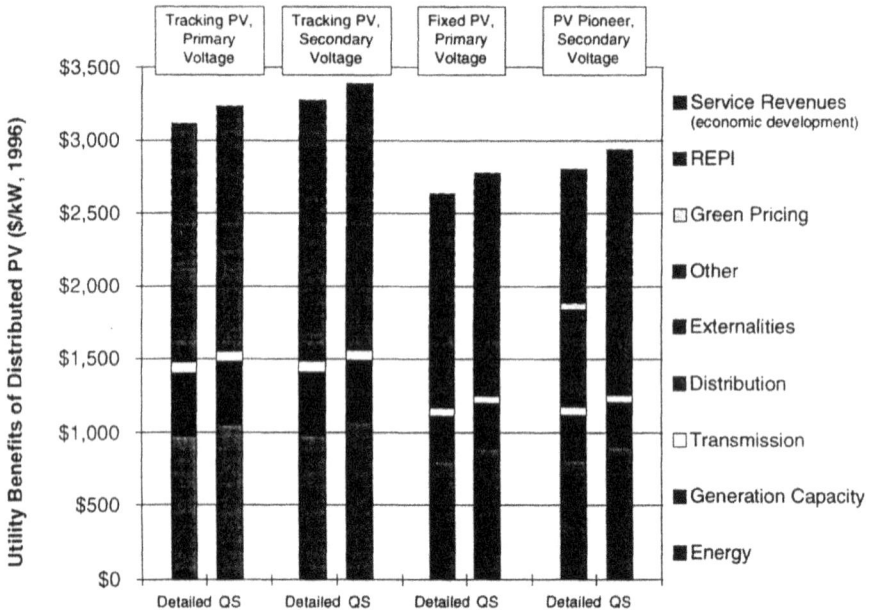

Wenger, Howard J., Thomas E. Hoff, and Jan Pepper (1996). *Photovoltaic Economics and Markets: The Sacramento Municipal Utility District as a Case Study*, p. 4-11.

---

[81] Wenger, Howard J., Thomas E. Hoff, and Jan Pepper (1996). *Photovoltaic Economics and Markets: The Sacramento Municipal Utility District as a Case Study*, pp. 4-10 and 4-11.

Figure 7-6. shows an example of a detailed cost-benefit analysis compared to results using the QuickScreen software. Utility-specific benefit categories not mentioned above are REPI (a US federal subsidy for renewable energies) and service revenues (net revenues from a PV manufacturing plant in the utility's service area). We observe that the results are almost coincidental, with QuickScreen only slightly overestimating the actual utility benefits from distributed power generation within a 5% error margin. The QuickScreen software allows electric utilities to estimate their monetary benefits from distributed power generation technologies in a quick and easy way. Although QuickScreen software was originally developed for the US territory, only minor modifications would be necessary to use this software package for assessing utility benefits in other countries[82].

It is interesting to note that in the chart's example, distributed benefits from various PV power technologies (tracking and fixed PV; primary and secondary voltage support) add up to almost 3,500 $/kWp. These values are utility-specific and have to be examined in a case-by-case assessment.

17.4.3 Cost-Benefit Analyses in Deregulated Power Markets

Several of the benefit-categories presented above refer to the deferral of generation capacity installations or transmission and distribution system upgrades. These benefits have been evaluated for vertically integrated utilities, owning and operating both generating units and the electric grid.

In a deregulated power market, unbundling will lead to the factual separation of generation, transmission, and distribution services. Critics of electric restructuring have argued that unbundling also means the end of integrated resource planning and least cost planning efforts.[83] Since formerly vertically integrated electric power companies are required to divide their firm into several independent subsections and, in case of further central ownership, warrant their individual operation and market participation, this unbundling process could diminish the incentive for system-wide least cost planning activities.

Investment in energy-saving efforts (demand side management, DSM) or the negawatt-strategy as advocated by Amory Lovins and the Rocky Mountain Institute

---

[82] Further information on QuickScreen is available from Christy Herig of the National Renewable Energy Laboratory, or from John Stevens of Sandia National Laboratories.

[83] Letendre, Steven, John Byrne, and Young-Doo Wang (1996). *The Distributed Utility Concept: Toward a Sustainable Electric Utility Sector*, p. 7-3.

could therefore become obsolete and lose their economic cost-effectiveness in the investment and cost calculations performed by unbundled entities. [84]

While some of the categories used above may be economically applicable for vertically integrated utilities only, other categories are valid for deregulated power systems also. Electric generators in a liberalized environment will face economic advantages through lower externalities; advantages in generating capacity; possible service revenues from solar equipment manufacturer cooperation; green pricing revenues; and risk mitigation.

17.5 Residential Customer Benefits from Distributed Electricity Generation

Distributed power generation is an asset for electric utilities to enhance the economics and reliability of their electricity system. Residential customers have indirect benefits from distributed generation. Small power generators running on renewable energies reduce environmental damages caused by electric power production.

Furthermore, distributed power technologies decrease the risk of stranded costs. Stranded costs for investment that has become uneconomic due to market changes have to be paid for by the electric customer. Electric restructuring stranded the investment costs for a multitude of central power stations. Small renewable power technologies have lower risks of becoming stranded investment, since environmentally friendly electric power can be marketed at higher than average market rates, and distributed generation capacity can be installed according to the actual energy demand for renewable resources.

Consumers may also profit directly from distributed generation. Net metering legislation allows customers to sell their electric current back to the grid. In the STELLA models of Part VIII, we will examine this issue in more detail. Net metering, of course, only works if residential customers own their small power systems.

In a restructured and deregulated market, residential power producers will also have the opportunity to market their electric power and negotiate electric power contracts with interested electricity consumers. In regulated and vertically integrated power markets, residential power producers have had the opportunity of generating and consuming their own power, but did not have the possibility to sell electricity to other residential customers. Selling their surplus power back to the

---

[84] Amory Lovin's alternative energy concept was first outlined in his book *Soft Energy Paths. Toward a Durable Peace*. New York: Harper & Row, Publishers, Inc., 1977. He generalized his negawatt-concept for negacars, negaliters, negamiles, etc. in *Factor Four* (German Edition: Ernst Ulrich von Weizsäcker, Lovins & Lovins, Faktor Vier. Doppelter Wohlstand – halbierter Naturverbrauch. München: Droemer Knaur, 1996).

utility has been the only option for SPPs in vertically integrated monopolistic power markets.

18. Micro-Grids: Electric Power Generation without a Central Grid System

18.1 Definition

The Pacific Energy Group has initiated research on micro-grid power systems. With this approach, economic and energetic benefits in electricity systems that are not interconnected with the central grid are assessed.

A micro-grid consists of several electric power generators that supply the demand of a group of customers. The micro-grid is electrically isolated, i.e. not connected to the public grid system. Micro-grids may consist of a limited number of residential homes that are interconnected with each other and form an isolated power system independent from the large utility grid.[85]

A micro-grid system can be owned and operated by a group of residential consumers as well as in customer-utility cooperation with the electric utility providing know-how, installation and maintenance service.

18.2 Components of Micro-Grid Systems

Micro-grids have to provide reliable electric power without relying on back-up capacity from the central grid system. It is therefore useful to combine both heat and electricity generation in an integral energy system. Cogeneration of electricity and heat is a crucial component of micro-grids. Fuel cell technologies based on natural gas or hydrogen provide space heating and hot water as well as electricity. For efficiency and environmental reasons, fuel cells are the most promising cogeneration technology.

Natural resources such as local hydropower, wind and solar generators are further parts of micro-grid systems. Microhydro turbines are a cost-effective option for providing electric services. The technology and economics of wind power generation have advanced remarkably over the last decades. Finally, photovoltaic modules can be integrated in micro-grid applications to provide power for peak-load demand.

Micro-grid systems are not burdened with the cost of a central grid system. To develop cost-effective micro-grids, demand-side management efforts have to be undertaken before determining and sizing necessary electric generation capacity.

---

[85] Hoff, Thomas E., Howard J. Wenger, Christy Herig, and Robert W. Shaw (1998). *A Micro-Grid With PV, Fuel Cells, and Energy Efficiency*, p. 2.

Figure 7-7. Energy Efficiency in a Central Power and in a Micro-Grid System

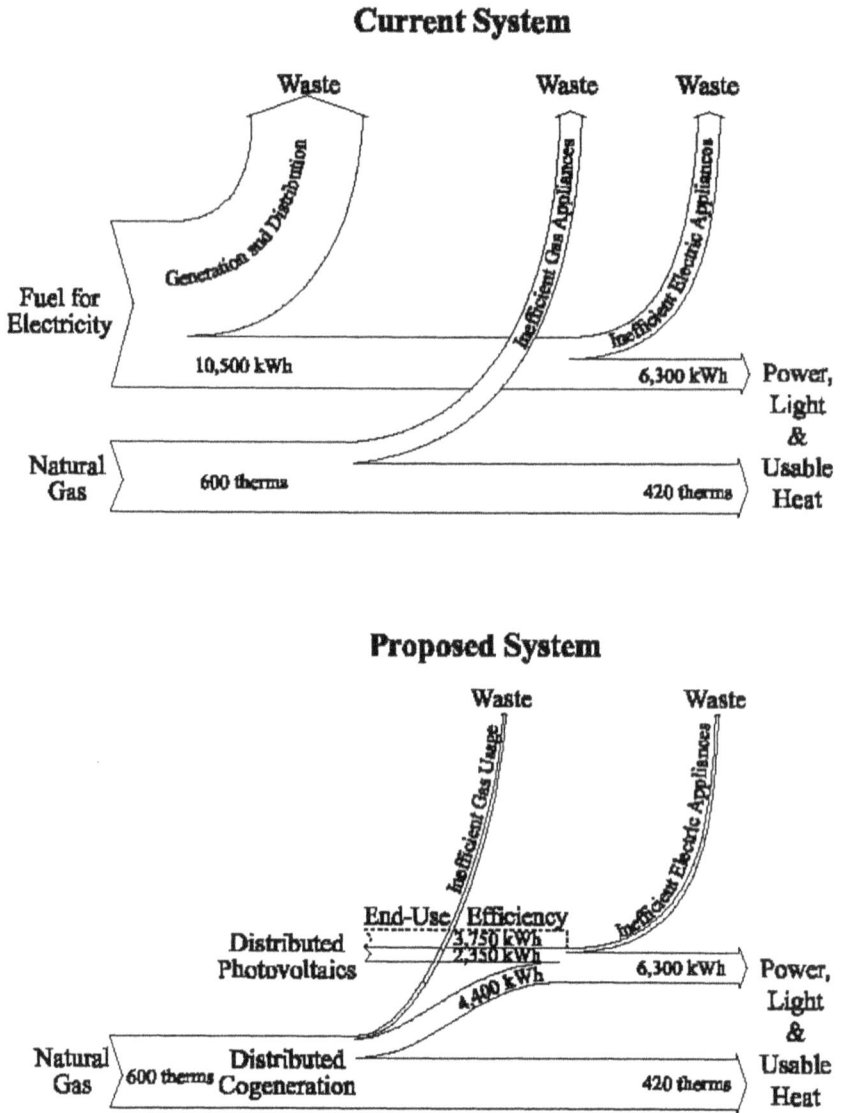

Current System

Proposed System

Hoff, Thomas E., Howard J. Wenger, Christy Herig, and Robert W. Shaw. (1998). *A Micro-Grid With PV, Fuel Cells, and Energy Efficiency*, p. 2.

As experiences in off-grid photovoltaic applications have demonstrated, it is generally cost-effective to cut down total system loads first by installing high-efficiency electric appliances or replacing electric appliances by other energy options (e.g. gas cooking and cooling instead of electric refrigeration and cooking). The cost-effectiveness of demand side management efforts will be shown in a STELLA-model of Part VIII.

18.3 Energy-Efficiency of Micro-Grids

The concept of micro-grids is based on the assumption that becoming independent from the central grid system would lead to total cost savings and increase the cost-effectiveness of small-scale renewable technologies, cogeneration, and energy-efficiency investment.

Figure 7-7. outlines another advantage of electric micro-grid systems. Micro-grids use energetic resources more effectively than central power systems. The upper part of the figure shows an example for a conventional system of central power generation and residential heat production with natural gas. Energy and money is wasted at the stages of central electricity generation, transmission and distribution, as well as through inefficient natural gas and electric appliances. The lower part of the figure outlines an alternative system based on cogeneration of heat and electricity and minimizing energetic and monetary losses. Distributed photovoltaics provide additional power and allow for the system's independence from central power generation. Improvement in energy-efficiency reduces the total electric power load.

18.4 Economics of Micro-Grid Systems

Whether a micro-grid system is cost-effective with central power generation largely depends on the costs for central power generation, transmission, and distribution. System-specific calculations have to determine the economics of micro-grid power production. From an energetic perspective, we have seen that micro-grid heat and electric power cogeneration can be advantageous. The cost-effectiveness of micro-grids depends on the cost of distributed power technologies for photovoltaics and fuel cells. In remote applications and for Third World countries, micro-grids may be cost-effective already. For industrialized countries with a high coverage of the central grid, electricity costs of micro-grid systems is not cost-effective at today's system costs for small-scale renewable generation technologies. The Pacific Energy Group estimates that current capital cost of photovoltaics is a factor 3 or 4 too high to make micro-grids cost-competitive with conventional power systems without any subsidies. Furthermore, there are no cost-effective residential cogeneration products available for residential consumer applications in the US (e.g. 2-kWp natural gas fuel cells).[86]

Mass-production of fuel cells and photovoltaic modules may improve the cost-effectiveness of micro-grid systems. In Germany, research and development for residential-size cogeneration systems has lead to commercially available gas-powered systems. Mass production is yet necessary to lower electric power generation costs. Current efforts of the automotive industry to develop small-scale hydrogen fuel cells for mass-production and their use in electric vehicles may speed the commercialization process. We will examine the economics of micro-grid components in Part VIII in more detail.

---

[86] Hoff, Thomas E., Howard J. Wenger, Christy Herig, and Robert W. Shaw (1998). *A Micro-Grid With PV, Fuel Cells, and Energy Efficiency*, p. 6.

PART VIII.

DYNAMIC MODELING - ASSESSING THE ECONOMICS OF PV POWER GENERATION IN REGULATED AND RESTRUCTURED ELECTRICITY MARKETS

19. STELLA Dynamic Modeling

The STELLA II software by High Performance Systems, Inc., USA, helps to understand complex real-world phenomena. The software uses an iconographic programming style and can be run on modern personal computers using Windows or Macintosh operating systems. In the context of this work, we will refer to the STELLA II software with the term STELLA.

The following section provides a short introduction into the basics of STELLA programming.

19.1 STELLA II Software

With STELLA II software, linear and dynamic systems can be modeled and examined. STELLA visualizes the interdependences between variables and offers the possibility of running sensitivity analyses to assess the impacts of varying selected variables to the complex system. The STELLA windows-bases programming surface shows the outline of the computer model, while the calculation results can be analyzed through tables and charts. STELLA also translates the windows-based programming operation into algebraic terms and mathematical formulas. All these features will be used in this chapter to provide a detailed insight into the models' outlines, calculations, and results.

19.2 STELLA Programming Elements

19.2.1 Stocks

Stocks function as reservoirs for numbers and empirical data. A stock may contain a group of elements, such as a number of people or an amount of money. For creating a stock, STELLA requires information about the initial stock value, that is the number of elements already in the stock at the beginning of the calculation. Stocks are symbolized by rectangles and can be named according to their contents or their function.

Stock 1

The contents of a stock can vary over time. A stock can be fed by inflows and be decreased by outflows. Since STELLA is a dynamic software program, we can perform calculations over a time period. This time period may literally be measured in months or years, but also symbolize calculation units without representing a time value at all. For our models, we shall often use the dynamic calculation process for analyzing systematic photovoltaic system cost decreases. In these cases, the PV system cost constitutes the stock.

19.2.2 Inflows and Outflows

Inflows add elements to the contents of a stock, while outflows decrease a stock's number of elements. Inflows and outflows are visualized by arrows flowing into or out of the stock. Little clouds indicate that the elements of the outflows and inflows go to and come from outside the closed model system.

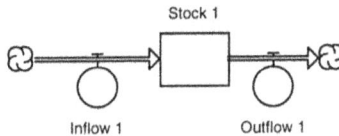

The value of the inflow can be set as a constant (e.g. inflow of one element each time period) or as a function. If a function is used, it is advantageous to use a separate symbol offered by STELLA.

Functions as well as any additional variables are visualized by a circle.

19.2.3 Variables

Variables determine the parameters of a model and the interaction between the model's components. Circles symbolize these parameters, and slim arrows are used to visualize the variables' interrelationships. The red arrow's head points to the direction of the variable's impact.

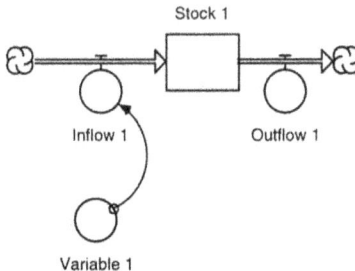

Variables can be constants or functions. In this way, equations can be created that form the algebraic basis of the dynamic model.

19.2.4 Conveyors

Sometimes it is useful to have elements wait in a stock for a determined time period. After this time period, which can be set by a variable, the element leaves the stock. This kind of stock is called a conveyor and visualized by a patterned rectangle.

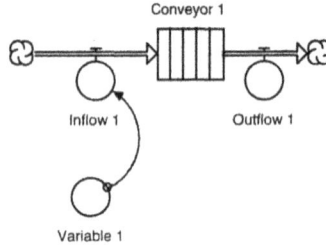

19.2.5 Ghosts

To enhance a model's transparency, it is sometimes advisable to use ghosts. Ghosts represent variables, stocks, or conveyors at another location. They are shaped the same way as their original, but can be identified by their dashed silhouette.[87]

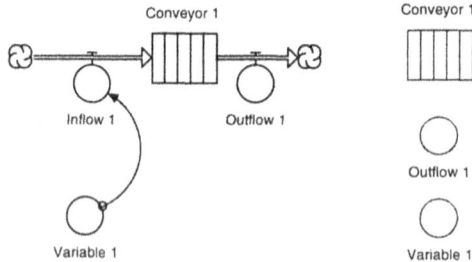

19.2.6 Feedback-Processes

STELLA's strength consists in visualizing and analyzing dynamic systems. Conventional calculation software may be advantageous for linear systems, but STELLA's versatility also includes dynamic systems featuring feedback processes such as positive and negative feedback, as shown in the next picture.

---

[87] Computer printouts sometime degrade the visibility of the dashed outline; since ghosts bear the same name as their original, they should however be easy to identify.

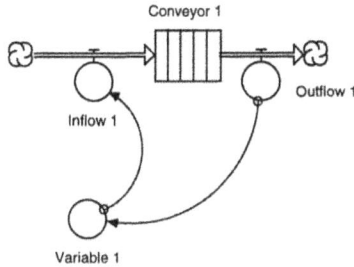

19.3 STELLA Analysis Elements

19.3.1 Tables

Tables represent the calculation's exact values. Tables are suited for determining a variable's concise mathematical value at a certain time or calculation time period.

Time	Conveyor 1	Inflow 1	Outflow 1	Variable 1
0	0	7	0	7
1	7	14	7	14
2	14	21	14	21
3	21	28	21	28
4	28	35	28	35
Final	35	42	35	42

19.3.2 Charts

Charts graphically illustrate the model calculation's results. Charts help to identify variable behavior; break-even points; oscillations; as well as patterned or chaotic system behavior.

The next picture shows a STELLA chart as an example. On the vertical axis, different values are printed for the three variables assessed. The horizontal axis represents time units or calculation steps.

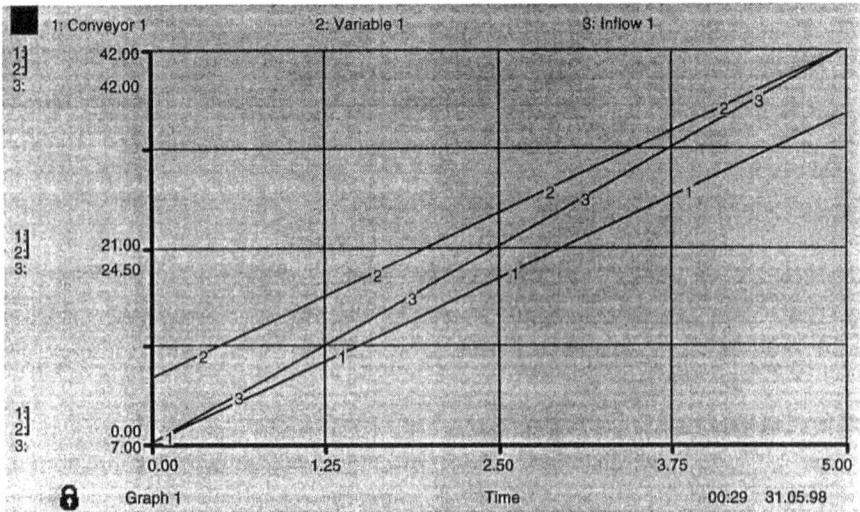

19.4 Calculation Methods

19.4.1 Time Runs

As indicated above, the models can be assessed by time runs based on either real time values (hours, days, years,...) or theoretical calculation units. The time or calculation units are usually visualized on the horizontal axis of the STELLA charts. For our models, we will mostly use time periods in analogy to calculation steps, with each calculation step reducing PV system costs by a certain value.

19.4.2 Sensitivity Runs

In order to analyze the systematic variation of a selected variable and impacts on other variables or the total model, sensitivity runs can be performed. The results of a sensitivity run can be examined through comparative printouts in either tables or charts.

The subsequent chapters will now derive various models for assessing the economics of PV power generation in various market scenarios from both the electric utilities' and the residential customers' perspective.

20. Dynamic Models

20.1 The Economics of PV Power Generation in a Regulated and Vertically Integrated Electricity Market

20.1.1 Utility PV Power Generation in a Regulated Environment

20.1.1.1 Deriving a Basic Model for Conventional Utility Cost Calculation

The paragraphs following will develop computer models to analytically discuss the economics of photovoltaic power generation. The models can be easily adapted to a variety of input factors and experimental conditions (DM or US$; solar radiation data for Germany and the USA; market conditions; etc.).

Figure 8-1. shows the modeling of PV system cost reductions. The initial stock (rectangle) is 15,000 DM/kWp and is systematically decreased (arrow with circle) by 500 DM/kWp through each STELLA calculation step.

Figure 8-1. Decrease of PV System Costs

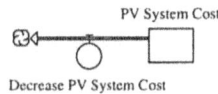

PV System Cost

Decrease PV System Cost

Figure 8-2. STELLA Chart of Decreasing PV System Cost Calculation

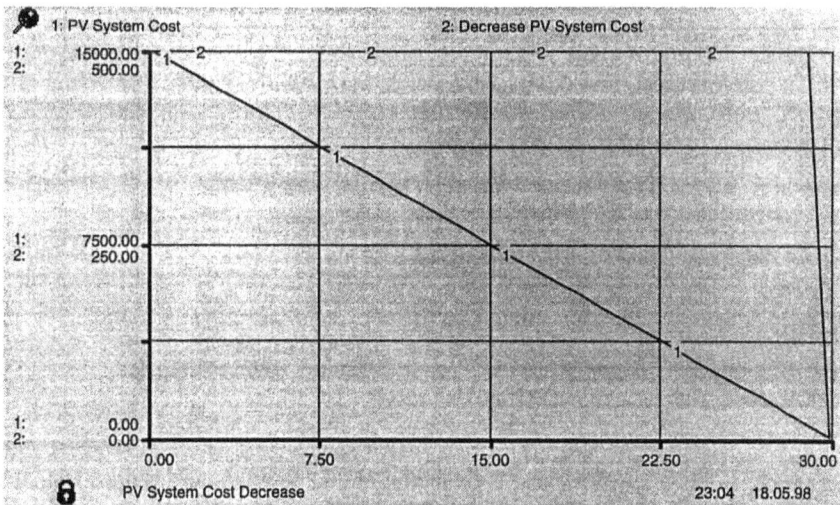

Figure 8-2. illustrates the STELLA calculation reducing the PV system cost by 500 DM/kWp steps. The vertical chart axis labels show the PV system price levels. The horizontal axis labels are somewhat more difficult to interpret, since STELLA by default calculates in time units. Here, of course, the time units represent calculation steps, with each step representing a 500 DM/kWp cost reduction. Therefore, the x-axis value of 0 represents a 15,000 DM/kWp cost level, the value of 15.00 a cost level of 7,500 DM/kWp, with zero cost being reached at the end of the horizontal chart axis. The STELLA charts following all build on this relationship and have to be interpreted accordingly.

Discount rates are an important input factor for investment calculations. If the influence of inflation is to be eliminated, real discount rates have to be calculated by subtracting the inflation rate from the nominal discount rate. This relationship is shown in Figure 8-3. An assumed nominal discount rate of 8% at an estimated inflation rate of 3.5% results in a real discount rate of 4.5%. These numbers are commonly used in the investor-owned electric utility industry.

Figure 8-3. Nominal and Real Discount Rate Calculation

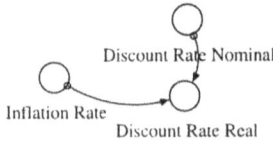

The regulated electricity industry usually uses levelized cost calculations for investment decision making. This approach estimates the annual costs that result from an investment, considering the company's discount rate (indicating the required profit margin) and investment time period. Levelized cost calculations can be performed for nominal and real values.

Figure 8-4. shows the input variables and analytical relationships for calculating the cost recovery factor (CRF) with the formula $CRF_{(nominal)} = (1+i)^{n*i} / (1+i)^{n-1}$

or with the input variables used in STELLA:

$CRF_{(nominal)} =$
((1+Discount_Rate_Nominal)^Time_Period*Discount_Rate_Nominal) /
((1+Discount_Rate_Nominal)^Time_Period-1)

Assuming a 20 years operation time period of a photovoltaic system and the discount rates mentioned above, the nominal CRF is 0.101852, and the real CRF

0.076876, respectively. Utility representatives often use a CRF value of 0.10 for rough estimates of PV power costs.[88]

Figure 8-4. Calculating the Cost Recovery Factor, CRF

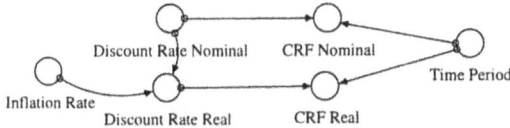

To calculate the annual fixed charge (AFC; Annuität), the cost recovery factor has to be multiplied with the total investment cost, as shown in Figure 8-5.

Figure 8-5. Calculating the Annual Fixed Charge (AFC)

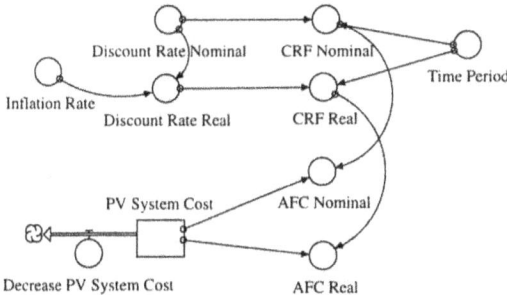

Balance of system costs (BOS) emerge from expenditures that occur with system operation, e.g. costs for maintenance; operation; inspection; necessary repairs; etc. In our model, we add BOS costs as a percentage to the annual fixed charge, hence discounting the BOS costs accordingly.

To determine the actual cost of energy in US$/kWh or DM/kWh, the total annual fixed charge (including annual BOS rate) has to be divided by the amount of electrical energy produced by the PV system (see Figure 8-6.). This is dependent on the solar radiation level of the region the PV system is installed in. Values vary

---

[88] Bzura, John J. (1997). *Basic Economics of Residential PV Systems*, p. 1.

from approximately 600 kWh/kWp in Northern Germany to 1,800 kWh/kWp in the Southern USA (see Table 8-3.). The following models and sensitivity analyses will address this item in more detail.

Figure 8-6. Calculating Cost of Energy (US$/kWh)

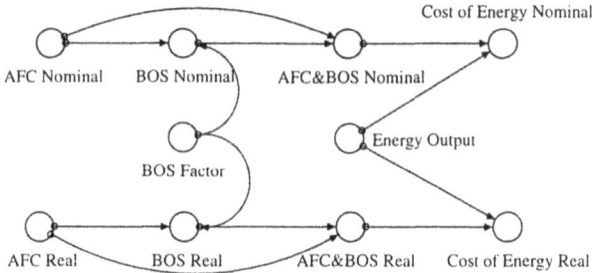

20.1.1.2 Model Assumptions

The basic model for conventional utility cost calculation is based on empirically valid assumptions on the variable values used.

1. PV System Costs

In respect to market studies of the German Fraunhofer Institute for Solar Systems on PV system costs examined in the 1000-Roofs-Program, today's lowest possible system cost is approximately at 15,000 DM/kWp installed capacity.[89] To examine further cost reductions, STELLA reduces total system costs in steps of 500 DM/kWp each calculation cycle (i.e. time period).

2. Balance of System Costs (BOS)

Studies on the economics of PV power generation that are available from electric utilities rather unsystematically deal with balance of system costs. These costs are frequently neglected, based on the argument that they are negligible in comparison to the cost of PV modules. PV module price would already constitute an insurmountable barrier against cost-competitiveness of solar electric power generation.

Nevertheless, we will address BOS costs. For our model, we assume the BOS costs to equal 1.5% of levelized annual system costs.

---

[89] Fraunhofer Institut für Solare Energiesysteme (1995). *1000-Dächer Meß- und Auswerteprogramm, Jahresjournal 1995*, p. 4.

Figure 8-7. Model for Conventional Utility Cost Calculation

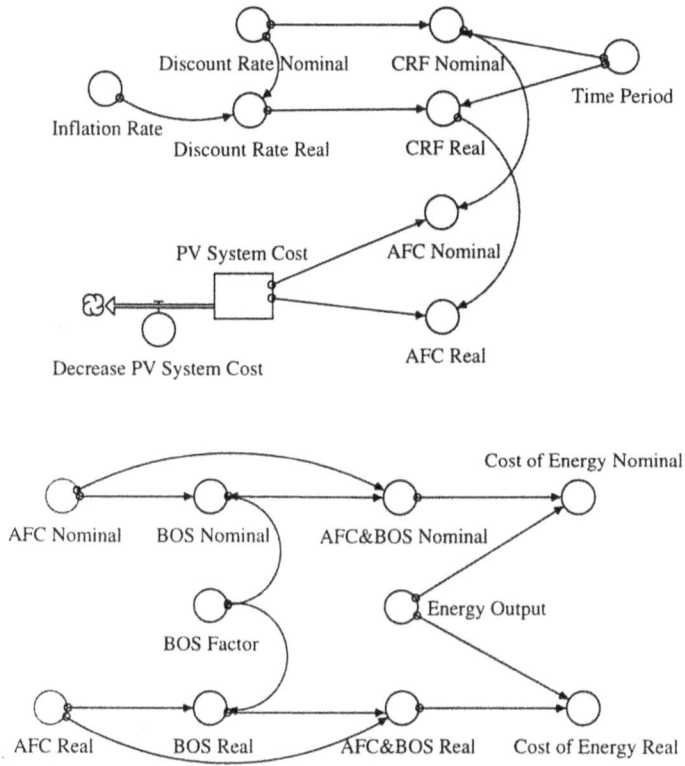

3. Discount and Inflation Rates

The model assumes a nominal discount rate of 8% to cover utility profits and inflation. This value is based on investment cost calculations performed by the German electric utilities Bayernwerk and RWE in cooperation with PV manufacturer Siemens Solar.[90] At an inflation rate of 3.5%, the real discount rate is 4.5%.

---

[90] Bayernwerk, Siemens, and RWE (1993). *Kostenentwicklung von Photovoltaik-Kraftwerken in Mitteleuropa*, p. 35.

4. System Life (Calculation Time Period)

In the model, system life coincides with the calculation time period, i.e. the utility does not intend to recover total investment costs before the scheduled operating time of the PV system expires.

State-of-the-art PV panels feature warranties up to 25 years.[91] Our model is based on a 20 years operating time of the PV system and a 20 years investment calculation period, respectively.

Figure 8-7. illustrates the model outline graphically as seen on the windows surface of the STELLA dynamic modeling software.

The model is based on the following mathematical equations and variables:

Stock:

PV_System_Cost(t) = PV_System_Cost(t - dt) + (- Decrease_PV_System_Cost) * dt

INIT PV_System_Cost = 15000

Outflow:

Decrease_PV_System_Cost = 500

Variables and Relationships:

AFC&BOS_Nominal = BOS_Nominal+AFC_Nominal

AFC&BOS_Real = AFC_Real+BOS_Real

AFC_Nominal = PV_System_Cost*CRF_Nominal

AFC_Real = PV_System_Cost*CRF_Real

BOS_Factor = 0.015

BOS_Nominal = AFC_Nominal*BOS_Factor

BOS_Real = AFC_Real*BOS_Factor

Cost_of_Energy_Nominal = AFC&BOS_Nominal/Energy_Output

Cost_of_Energy_Real = AFC&BOS_Real/Energy_Output

---

[91] Siemens Solar offers this warranty for its latest PV modules on a determined minimum energy output.

CRF_Nominal =
((1+Discount_Rate_Nominal)^Time_Period*Discount_Rate_Nominal) /
((1+Discount_Rate_Nominal)^Time_Period-1)

CRF_Real = ((1+Discount_Rate_Real)^Time_Period*Discount_Rate_Real)
/ ((1+Discount_Rate_Real)^Time_Period-1)

Discount_Rate_Nominal = 0.08

Discount_Rate_Real = Discount_Rate_Nominal-Inflation_Rate

Energy_Output = 925

Inflation_Rate = 0.035

Time_Period = 20

20.1.1.3 Sensitivity Analyses

We will now use sensitivity analyses to assess the systematic variation of selected variables on the model results. Nominal and real cost of energy will function as the dependent variable when running sensitivity analyses on PV system cost; balance of system cost; the discount rates used; the time period of the calculation; and energy output data.

20.1.1.3.1 Sensitivity Analysis on PV System Cost

Table 8-1. illustrates the effect of system cost reductions on the cost of energy. Installed PV systems currently available are at 15,000 DM/kWp. Including 1.5% BOS cost on the annual fixed charge, the calculated cost of energy is at 1.68 DM/kWh nominal or 1.27 DM/kWh in real prices, respectively.

The table also shows that it is highly important to distinguish between nominal and real values. Real values consider the inflation rate and should be used for investment decision making, since the operation and investment time frame for a PV system is rather long. Advocates for and against renewable power generation sometimes misuse the instrument of real and nominal values to stress their arguments. Using nominal prices, higher cost levels can be calculated to advocate a higher full cost rate contributed to small PV power producers. Real prices, on the other hand, can be used to indicate the economic feasibility of PV power generation, reaching cost-competitiveness levels sooner than anticipated by critics.

For comparisons, it is important to stick to one method. In the following assessments, we will usually use real prices in order to address inflation. For some sensitivity analyses, nominal values are necessary and cannot be avoided.

Table 8-1. Sensitivity Results on PV System Cost Variation

Time	PV System Cost DM/kWp	BOS Nominal DM/a/kWp	BOS Real DM/a/kWp	Cost of Energy Nominal DM/kWh	Cost of Energy Real DM/kWh
0	15,000.00	22.92	17.30	1.68	1.27
1	14,500.00	22.15	16.72	1.62	1.22
2	14,000.00	21.39	16.14	1.56	1.18
3	13,500.00	20.63	15.57	1.51	1.14
4	13,000.00	19.86	14.99	1.45	1.10
5	12,500.00	19.10	14.41	1.40	1.05
6	12,000.00	18.33	13.84	1.34	1.01
7	11,500.00	17.57	13.26	1.29	0.97
8	11,000.00	16.81	12.68	1.23	0.93
9	10,500.00	16.04	12.11	1.17	0.89
10	10,000.00	15.28	11.53	1.12	0.84
11	9,500.00	14.51	10.95	1.06	0.80
12	9,000.00	13.75	10.38	1.01	0.76
13	8,500.00	12.99	9.80	0.95	0.72
14	8,000.00	12.22	9.23	0.89	0.67
15	7,500.00	11.46	8.65	0.84	0.63
16	7,000.00	10.69	8.07	0.78	0.59
17	6,500.00	9.93	7.50	0.73	0.55
18	6,000.00	9.17	6.92	0.67	0.51
19	5,500.00	8.40	6.34	0.61	0.46
20	5,000.00	7.64	5.77	0.56	0.42
21	4,500.00	6.88	5.19	0.50	0.38
22	4,000.00	6.11	4.61	0.45	0.34
23	3,500.00	5.35	4.04	0.39	0.30
24	3,000.00	4.58	3.46	0.34	0.25
25	2,500.00	3.82	2.88	0.28	0.21
26	2,000.00	3.06	2.31	0.22	0.17

27	1,500.00	2.29	1.73	0.17	0.13
28	1,000.00	1.53	1.15	0.11	0.08
29	500	0.76	0.58	0.06	0.04
Final	0	0.00	0.00	0.00	0.00

Table 8-1. also shows that significant cost reductions are necessary for residential PV technologies to reach cost-effectiveness levels. In some German regions, electric utility rates reach up to 0.25-0.30 DM/kWh for residential customers (see Table 8-7.). With rates equal costs and hence underestimating utility profit margins, this would require PV system costs to drop to 3,500 to 3,000 DM/kWp in real values. Cost-effectiveness with fossil fuel technologies ranging best at 0.08 to 0.10 DM/kWh[92] would require PV system costs to drop to approximately 1,000-1,200 DM/kWp under the model assumptions.

20.1.1.3.2 Sensitivity Analysis on Balance of System Cost

Figure 8-8. Sensitivity Analysis on BOS Cost

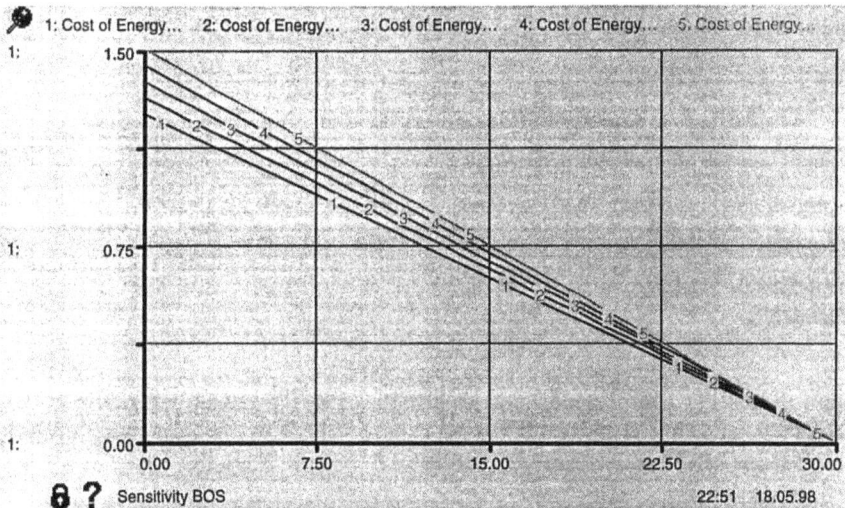

---

[92] as estimated by Vereinigung Deutscher Elektrizitätswerke (VDEW), *Strom-Daten Februar 1997*. Kapitel 12: Brennstoff- und Erzeugungskosten. Frankfurt, VDEW e.V.: 1997.

Increasing BOS cost could have considerable impact on the costs of PV power generation. Sensitivity runs for BOS factors of 0%; 5%; 10%; 15%; and 20%, result in real energy costs of 1.25 DM/kWh; 1.31 DM/kWh; 1.37 DM/kWh; 1.43 DM/kWh; and 1,50 DM/kWh (see Figure 8-8., lines 1-5). Approaching low levels of total system costs decreases the significance of BOS costs. In fact, there may be further BOS cost savings through learning curve effects in system maintenance and operation reliability. A certain level of BOS costs will probably remain and add to the initial investment cost.

20.1.1.3.3 Sensitivity Analysis on Discount Rate

The discount rate used for utility investment calculation substantially influences cost-levels of PV power generation. RWE and Bayernwerk use a nominal discount rate of 8% at an estimate inflation rate of 3.5%, thus calculating a real discount rate of 4.5%.[93] US investor-owned utilities such as New England Electric Services also use nominal discount rates of 8%.[94] For municipal utilities, lower nominal discount rates are commonly used. The Sacramento Municipal Utility District calculates with a nominal discount rate of 6.6%, which leads to a real discount rate of 3.1% at an estimated inflation rate of 3.5%.[95]

Our next model sensitivity analysis assesses various nominal discount rates at a constant 3.5% inflation rate. The sensitivity parameters are shown in Table 8-2. and illustrated in Figure 8-9.

Table 8-2. Sensitivity Parameters for Discount Rate Variation

Chart Line # (in Figure 8-9.)	1	2	3	4	5
Nominal Discount Rate	4%	6%	8%	10%	12%
Real Discount Rate	0.5%	2.5%	4.5%	6.5%	8.5%
Real Cost of Energy (DM/kWh)	0.87	1.06	1.27	1.49	1.74

With decreasing real discount rates as low as 0.5%, PV power generation at 0.87 DM/kWh is feasible at today's PV system cost levels. However, it is rather unlikely

[93] Bayernwerk, Siemens, RWE (1993). *Kostenentwicklung von Photovoltaik-Kraftwerken in Mitteleuropa*, p. 35.

[94] Bzura, John J. (1997). *Basic Economics of Residential PV Systems*, p. 1.

[95] Wenger, Howard J. (1997). *PV Cost Calculation*. Personal Email Conversation, July 18th, 1997, Sacramento Municipal Utility District, Sacramento, California.

that commercial utilities will use real discount rates at this level. On the other hand, costs as high as 1.74 DM/kWh can be calculated for real discount rates as high as 8.5%. The actual discount rate used is dependent on the type of electric utility (public or investor-owned, municipal, etc.) and its economic targets of investment recovery.

Figure 8-9. Sensitivity Analysis on Discount Rate

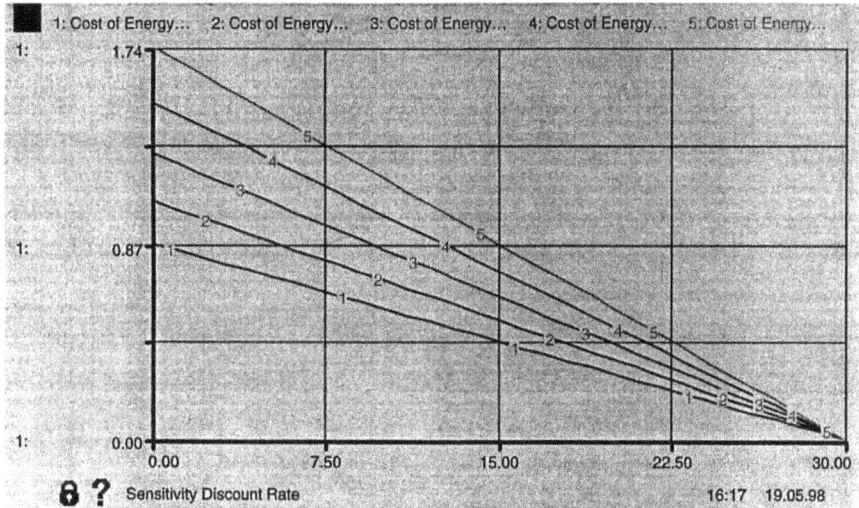

20.1.1.3.4 Sensitivity Analysis on the Calculation Time Period

Our model assumes coinciding time periods of PV system operation and investment calculation. PV modules have operated for decades in space and terrestrial applications providing reliable power for space satellites and remote power systems. Dependent on the PV cell technology used, these systems have experienced various levels of degradation in electric power output. Since no mechanical parts are involved in PV panels, they have rarely experienced defects (except for destruction through vandalism).[96]

Our model calculates real cost of energy for system operation and investment time periods of 10; 20; 30; 40; and 50 years (see lines 1 through 5, Figure 8-10.). The analysis evaluates real cost of energy reaching up to 2.08 DM/kWh for the 10-years

---

[96] For a detailed description of PV system performance for selected applications, see *Home Power Magazine*, Ashland, Oregon, USA, any issue.

calculation time period, 1.27 DM/kWh for 20 years, 1.01 DM/kWh for 30 years, 0.89 DM/kWh for 40 years, and 0.83 DM/kWh for the 50 years timeframe. It is interesting to note that short calculation periods seem to have a significant impact on the level of real energy cost, while extending calculation periods to 30, 40, or 50 years results in rather low additional cost reductions only.

Figure 8-10. Sensitivity Analysis on Time Period of Investment Calculation

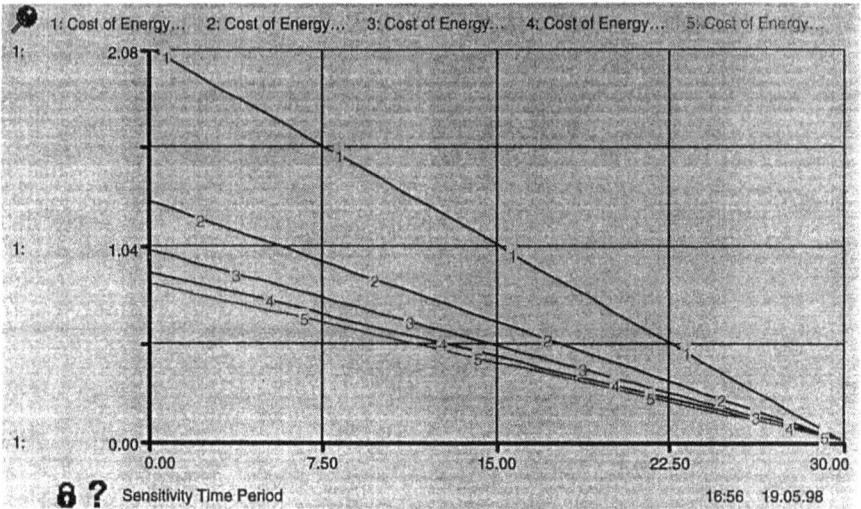

20.1.1.3.5 Sensitivity Analysis on Energy Output Data

The electric energy produced by a photovoltaic system is dependent on local insolation levels. For German regions, annual energy output varies between approximately 500 and 1000 kWh/kWp due to differences in average annual solar radiation. In Southern States of the US, energy output can reach as high as 1,800 kWh/kWp (see Table 8-3.).

This is why some scientists advocate the installation of large solar power stations in southern countries with high solar radiation and transmitting the generated electricity through high-energy power lines to industrialized countries.[97]

---

[97] Langniß, Luther, Nitsch and Wiemken (1997). *Strategien für eine nachhaltige Energieversorgung – Ein solares Langfristszenario für Deutschland*, p. 27.

**Table 8-3. Average Annual Energy Output due to Regional Differences
in Solar Radiation (Selected Systems)**

Country / State	Average Annual Energy Output in kWh/kWp 1995	Maximum Annual Energy Output in kWh/kWp 1995
Germany		
Baden-Württemberg	777	827
Bayern	699	814
Berlin	677	775
Brandenburg	704	857
Bremen	589	731
Hamburg	591	744
Hessen	700	793
Mecklenburg-Vorpommern	716	718
Niedersachsen	686	844
Nordrhein-Westfalen	690	905
Rheinland-Pfalz	731	806
Saarland	718	683
Sachsen	658	784
Sachsen-Anhalt	758	950
Schleswig-Holstein	689	892
Thüringen	655	760
Total Average	*689*	*805*
USA	*Estimated Average Energy Output*	
New England (NEES)	1,150	
California (SMUD)	1,810	

Figure 8-11. illustrates the model sensitivity analysis on various energy output
factors. Assumed energy output factors are 600 kWh/kWp (line 1; compares to low
German insolation levels); 900 kWh/kWp (line 2; compares to high German

insolation levels); 1,200 kWh/kWp (line 3; compares to New England insolation levels); 1,500 kWh/kWp (line 4); and 1,800 kWh/kWp (line 5; compares to California insolation levels). At current PV system costs, these values result in real cost of energy of 1.95 DM/kWh; 1.30 DM/kWh; 0.98 DM/kWh; 0.78 DM/kWh; and 0.65 DM/kWh, respectively.

Electric power output is dependent on regional solar radiation conditions only. In order to increase electric energy output, the PV systems have to be situated in regions of higher insolation, or technical efficiency increases have to lead to higher energy output per square meter of solar cell surface.

Figure 8-11. Sensitivity Analysis on Electric Energy Output

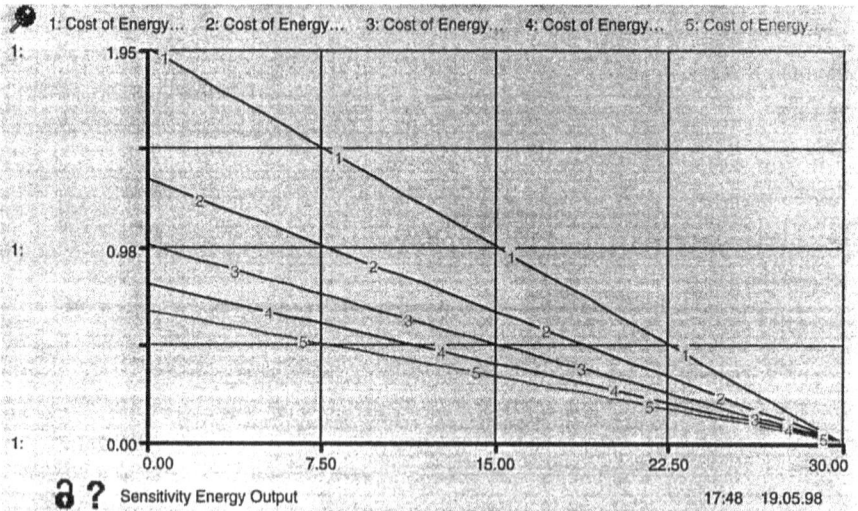

20.1.1.4 Scenarios

The following scenarios assess typical model assumptions valid for the US and for Germany.

The first two model scenarios examine the economics of grid-connected photovoltaic systems for an investor-owned electric utility in New England and for a municipal utility in California. The first case struggles with high discount rates for the investor-owned utility as well as comparatively low insolation levels, while the second case analyses a municipal utility with thus lower discount rates and high solar radiation in California. The results will therefore give the range of today's photovoltaic power generation cost in form of a best/worst-case dichotomy for the United States.

20.1.1.4.1 New England Electric Services (NEES) Case Model

In 1986, the investor-owned electric utility New England Electric Services (NEES) started to assess distributed power generation with residential grid-connected PV systems in Gardener, Massachusetts. This was the first New England Electric PV project. It consisted of 30 homes with 2-kWp PV systems and five commercial sites with system capacity ranging between 1.8 and 7.3 kWp. The systems have provided reliable solar electricity since 1986 with only minor technical failures, sometimes due to vandalism. Technical defects could be easily repaired in most cases.[98]

In order to set up a model for NEES PV cost calculation, we have to rely on current PV system cost data. The Gardener systems have been installed more than a decade ago and PV systems have experienced remarkable cost reductions since then. The Gardener project was designed for assessing technical reliability of grid-connected PV power generation, not for analyzing economic data.

Our NEES case model assumes total system costs (including parts, labor, and BOS costs) of 6,030 $/kWp. NEES' nominal discount rate is 8% for being an investor-owned electric power company. PV system energy output is estimated to be 1,150 kWh/kWp for New England solar radiation data. NEES assumes a 20-year system life and investment calculation period.[99]

Our STELLA model is then based on the following variables:

Stock:

PV_System_Cost(t) = PV_System_Cost(t - dt) + (- Decrease_PV_System_Cost) * dt

INIT PV_System_Cost = 6030

Outflow:

Decrease_PV_System_Cost = 500

---

[98] Bzura, John J. (1994). *Photovoltaic Research and Demonstration Activities at New England Electric*, pp. 1-6.

[99] Bzura, John J. (1997). *Basic Economics of Residential PV Systems*, p. 1.

Variables and Relationships:

AFC&BOS_Nominal = BOS_Nominal+AFC_Nominal

AFC&BOS_Real = AFC_Real+BOS_Real

AFC_Nominal = PV_System_Cost*CRF_Nominal

AFC_Real = PV_System_Cost*CRF_Real

BOS_Factor = 0

BOS_Nominal = AFC_Nominal*BOS_Factor

BOS_Real = AFC_Real*BOS_Factor

Cost_of_Energy_Nominal = AFC&BOS_Nominal/Energy_Output

Cost_of_Energy_Real = AFC&BOS_Real/Energy_Output

CRF_Nominal =
((1+Discount_Rate_Nominal)^Time_Period*Discount_Rate_Nominal)
/ ((1+Discount_Rate_Nominal)^Time_Period-1)

CRF_Real = ((1+Discount_Rate_Real)^Time_Period*Discount_Rate_Real)
/ ((1+Discount_Rate_Real)^Time_Period-1)

Discount_Rate_Nominal = 0.08

Discount_Rate_Real = Discount_Rate_Nominal-Inflation_Rate

Energy_Output = 1150

Inflation_Rate = 0.035

Time_Period = 20

Table 8-4. shows the model results on the assumptions indicated above. New England Electric Services calculates real energy costs of 0.40 $/kWh at today's system cost levels. Whether this cost level is cost-competitive in today's market environment can be derived from an overview of US retail price levels for residential electric customers.

Table 8-4. NEES Case Model Cost Analysis

Time	PV System Cost $/kWp	Cost of Energy Nominal $/kWh	Cost of Energy Real $/kWh
0	6,030.00	0.53	0.40
1	5,530.00	0.49	0.37
2	5,030.00	0.45	0.34

3	4,530.00	0.40	0.30
4	4,030.00	0.36	0.27
5	3,530.00	0.31	0.24
6	3,030.00	0.27	0.20
7	2,530.00	0.22	0.17
8	2,030.00	0.18	0.14
9	1,530.00	0.14	0.10
10	1,030.00	0.09	0.07
11	530	0.05	0.04
12	30	0.00	0.00
Final	0	0.00	0.00

The US map in Figure 8-12. visualizes the broad range of utility revenues from sales to electric retail customers by States. Retail revenues are between 4 and 10 cents/kWh. The red color indicates States where electric retail prices are highest.

The electricity price level is important for the economics of photovoltaic power generation. States such as California, Hawaii, and Arizona have advantageous opportunities for photovoltaics with abundant insolation and high electric retail prices.

Although the New England States in the Northeast feature lower solar radiation, their price level for electricity is comparitively high and thus advantageous for costly renewable power generation technologies.

However, it soon becomes clear that photovoltaic electric costs of real 0.40 $/kWh as determined by our model above are yet far from cost-competitiveness, regarding current average utility revenues of approximately 0.10 $/kWh for the New England States. With New England Electric Services' assumption on PV costs and investment calculation variables, PV power generation is not cost-competitive in NEES' market today.

PV system costs (including BOS) would have to decline to 1,500 $/kWp to even recover investment costs through market revenues only, not regarding the alternative utility investment option of fossil fuel power generation.

Figure 8-12. Average Utility Revenue from Electricity Sales to All Retail Consumers by State, in 1995 Cents per Kilowatthour

U.S. Average = 6.9

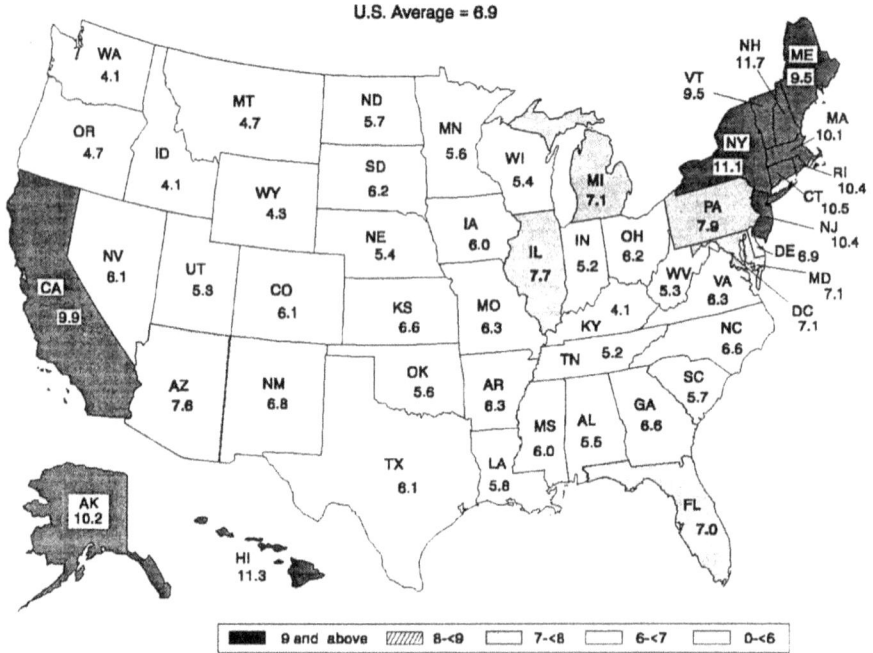

Energy Information Administration EIA / US Department of Energy DOE (1996). *The Changing Structure of the Electric Power Industry: An Update*, p. 36.

20.1.1.4.2 Sacramento Municipal Utility District (SMUD) Case Model

The Sacramento Municipal Utility District (SMUD) is a municipal utility and has lower cost of funds and thus lower discount rates than investor-owned utilities. SMUD usually calculates with a nominal discount rate of 6.6%. As for economic calculations on its PV systems, SMUD assumes a system operation and investment calculation time period of 30 years. PV electricity production is higher in California than in the New England States, and SMUD has experienced electric energy output of 1,810 kWh/kWp.[100]

---

[100] Wenger, Howard J. (1997). *PV Cost Calculation*. Personal Email Conversation, July 18th, 1997, Sacramento Municipal Utility District, Sacramento, California.

With these assumptions, the STELLA model for the SMUD case runs on the following variables:

Stock:

PV_System_Cost(t) = PV_System_Cost(t - dt) + (- Decrease_PV_System_Cost) * dt

INIT PV_System_Cost = 6030

Outflow:

Decrease_PV_System_Cost = 500

Variables and Relationships:

AFC&BOS_Nominal = BOS_Nominal+AFC_Nominal

AFC&BOS_Real = AFC_Real+BOS_Real

AFC_Nominal = PV_System_Cost*CRF_Nominal

AFC_Real = PV_System_Cost*CRF_Real

BOS_Factor = 0

BOS_Nominal = AFC_Nominal*BOS_Factor

BOS_Real = AFC_Real*BOS_Factor

Cost_of_Energy_Nominal = AFC&BOS_Nominal/Energy_Output

Cost_of_Energy_Real = AFC&BOS_Real/Energy_Output

CRF_Nominal =
((1+Discount_Rate_Nominal)^Time_Period*Discount_Rate_Nominal)
/ ((1+Discount_Rate_Nominal)^Time_Period-1)

CRF_Real = ((1+Discount_Rate_Real)^Time_Period*Discount_Rate_Real)
/ ((1+Discount_Rate_Real)^Time_Period-1)

Discount_Rate_Nominal = 0.066

Discount_Rate_Real = Discount_Rate_Nominal-Inflation_Rate

Energy_Output = 1810

Inflation_Rate = 0.035

Time_Period = 30

The rather favorable conditions valid for SMUD, such as high insolation levels in California and a comparatively low discount rate for its municipal utility status, together with SMUD's optimistic assumption of a 30-years operating and calculation timeframe result in significantly lower costs for photovoltaic power. Assuming NEES system costs from the previous model, the SMUD model calculates real cost of energy as low as 0.17 $/kWh (see Table 8-5.).

Additionally, SMUD has embarked on a strategic path to systematically reduce PV system costs. This approach will be assessed later in more detail. The program has been initiated in 1993 and since then considerably influenced SMUD's PV system costs in a yet regulated environment of vertically integrated electric utilities. We will use SMUD's actual system cost to more exactly determine costs for PV power generation feasible in California.

Through its Sustained Orderly Development Strategy based on long-term contracts, SMUD reduced total system costs (including mounting and operating cost) to 4,530 $/kWp in 1998.[101] This cost level results in real PV energy costs of 0.13 $/kWh which is only 3 cents above average electricity prices in the Californian retail market before electric restructuring (see map in Figure 8-12.). However, this level is still above retail market rates and even less attractive compared to conventional power generation alternatives.

Table 8-5. SMUD Case Model Cost Analysis

Time	PV System Cost $/kWp	Cost of Energy Nominal $/kWh	Cost of Energy Real $/kWh
0	6,030.00	0.26	0.17
1	5,530.00	0.24	0.16
2	5,030.00	0.22	0.14
3	4,530.00	0.19	0.13
4	4,030.00	0.17	0.12
5	3,530.00	0.15	0.10
6	3,030.00	0.13	0.09
7	2,530.00	0.11	0.07
8	2,030.00	0.09	0.06
9	1,530.00	0.07	0.04

[101] Wenger, Howard J., Thomas E. Hoff, and Jan Pepper (1996). *Photovoltaic Economics and Markets: The Sacramento Municipal Utility District as a Case Study*, p. 5-3.

10	1,030.00	0.04	0.03
11	530	0.02	0.02
12	30	0.00	0.00
Final	0	0.00	0.00

20.1.1.4.3 High-Cost Scenario Germany

To model photovoltaic electric generation cost for Germany, we first set up a high-cost scenario assuming investment decision parameters for an investor-owned electric utility with a high nominal discount rate. The utility shall be situated in a region with comparatively low solar radiation.

We therefore rate solar energy output at 500 kWh/kWp and a nominal discount factor of 8% at a constant inflation rate of 3.5%, as generally used in our models. PV system prices start at 15,000 DM/kWp, which is the minimum cost level as found in the Fraunhofer Institute studies mentioned above. Our PV systems expect a 20-year lifetime coinciding with investment period duration. This period is derived from the assumptions of the investor-owned utility NEES analyzed above.

These assumptions result in the set-up parameters for our STELLA model as follows:

Stock:

PV_System_Cost(t) = PV_System_Cost(t - dt) + (- Decrease_PV_System_Cost) * dt

INIT PV_System_Cost = 15000

Outflow:

Decrease_PV_System_Cost = 500

Variables and Relationships:

AFC&BOS_Nominal = BOS_Nominal+AFC_Nominal

AFC&BOS_Real = AFC_Real+BOS_Real

AFC_Nominal = PV_System_Cost*CRF_Nominal

AFC_Real = PV_System_Cost*CRF_Real

BOS_Factor = 0

BOS_Nominal = AFC_Nominal*BOS_Factor

BOS_Real = AFC_Real*BOS_Factor

Cost_of_Energy_Nominal = AFC&BOS_Nominal/Energy_Output

Cost_of_Energy_Real = AFC&BOS_Real/Energy_Output

CRF_Nominal =
((1+Discount_Rate_Nominal)^Time_Period*Discount_Rate_Nominal)
/ ((1+Discount_Rate_Nominal)^Time_Period-1)

CRF_Real = ((1+Discount_Rate_Real)^Time_Period*Discount_Rate_Real)
/ ((1+Discount_Rate_Real)^Time_Period-1)

Discount_Rate_Nominal = 0.08

Discount_Rate_Real = Discount_Rate_Nominal-Inflation_Rate

Energy_Output = 500

Inflation_Rate = 0.035

Time_Period = 20

The model results indicate that for the German high-cost scenario, photovoltaic power generation is yet far from being cost-competitive with electricity costs reaching 2.31 DM/kWh assuming current PV system cost levels (see Table 8-6.). PV system costs would have to drop to 1,500 DM/kWh to reach average retail price levels for German residential customers that average 0.25 DM/kWh (excluding VAT)[102]. To be cost-competitive with conventional power generation, PV system costs would even have to decline to as low as 500 DM/kWp, resulting in real electricity costs of 0.08 DM/kWh. Table 8-7. illustrates regional and average German utility revenues from power sales to residential and industrial customers.

Table 8-6. German High-Cost Scenario Cost Analysis

Time	PV System Cost DM/kWp	Cost of Energy Nominal DM/kWh	Cost of Energy Real DM/kWh
0	15,000.00	3.06	2.31
1	14,500.00	2.95	2.23

---

[102] Vereinigung Deutscher Elektrizitätswerke (1997), *Jahreserhebung bei Elektrizitätsversorgungsunternehmen für das Jahr 1996*, Tabelle 1.1 Stromabsatz und Erlöse.

2	14,000.00	2.85	2.15
3	13,500.00	2.75	2.08
4	13,000.00	2.65	2.00
5	12,500.00	2.55	1.92
6	12,000.00	2.44	1.85
7	11,500.00	2.34	1.77
8	11,000.00	2.24	1.69
9	10,500.00	2.14	1.61
10	10,000.00	2.04	1.54
11	9,500.00	1.94	1.46
12	9,000.00	1.83	1.38
13	8,500.00	1.73	1.31
14	8,000.00	1.63	1.23
15	7,500.00	1.53	1.15
16	7,000.00	1.43	1.08
17	6,500.00	1.32	1.00
18	6,000.00	1.22	0.92
19	5,500.00	1.12	0.85
20	5,000.00	1.02	0.77
21	4,500.00	0.92	0.69
22	4,000.00	0.81	0.62
23	3,500.00	0.71	0.54
24	3,000.00	0.61	0.46
25	2,500.00	0.51	0.38
26	2,000.00	0.41	0.31
27	1,500.00	0.31	0.23
28	1,000.00	0.20	0.15
29	500	0.10	0.08
Final	0	0.00	0.00

Table 8-7. Average Revenues for Electric Utilities from German Power Customers (electric rates listed without VAT)

German States	Average Revenues from Industrial Customers in DPf/kWh (1996)	Average Revenues from Residential Customers in DPf/kWh (1996)
Baden-Württemberg	16.37	24.23
Bayern	15.00	25.25
Berlin	19.33	29.25
Brandenburg	15.70	27.42
Bremen	13.42	27.67
Hamburg	12.43	27.44
Hessen	14.84	24.05
Mecklenburg-Vorpommern	20.07	27.55
Niedersachsen	13.18	22.69
Nordrhein-Westfalen	12.50	22.74
Rheinland-Pfalz	13.29	22.48
Saarland	14.44	24.52
Sachsen	17.95	26.80
Sachsen-Anhalt	15.50	27.70
Schleswig-Holstein	14.78	24.52
Thüringen	18.16	28.51
German Average Revenue	*14.37*	*24.51*

20.1.1.4.4 Low-Cost Scenario Germany

The low-cost scenario assumes investment decision parameters for a municipal electric utility with a lower nominal discount rate than an investor-owned utility. In order to create favorable model conditions, the municipal utility shall be situated in a region with comparatively high solar radiation.

Our German low-cost scenario model assumes solar energy output to equal 1000 kWh/kWp and a nominal discount factor of 6.6% (in analogy to the municipal utility SMUD) at a constant inflation rate of 3.5%. PV system prices again start at 15,000 DM/kWp. Our model assumes a PV system lifetime and an investment analysis timeframe of 30 years each.

The STELLA model then runs on the following variables:

Stock:

PV_System_Cost(t) = PV_System_Cost(t - dt) + (- Decrease_PV_System_Cost) * dt

INIT PV_System_Cost = 15000

Outflow:

Decrease_PV_System_Cost = 500

Variables and Relationships:

AFC&BOS_Nominal = BOS_Nominal+AFC_Nominal

AFC&BOS_Real = AFC_Real+BOS_Real

AFC_Nominal = PV_System_Cost*CRF_Nominal

AFC_Real = PV_System_Cost*CRF_Real

BOS_Factor = 0

BOS_Nominal = AFC_Nominal*BOS_Factor

BOS_Real = AFC_Real*BOS_Factor

Cost_of_Energy_Nominal = AFC&BOS_Nominal/Energy_Output

Cost_of_Energy_Real = AFC&BOS_Real/Energy_Output

CRF_Nominal = ((1+Discount_Rate_Nominal)^Time_Period*Discount_Rate_Nominal) / ((1+Discount_Rate_Nominal)^Time_Period-1)

CRF_Real = ((1+Discount_Rate_Real)^Time_Period*Discount_Rate_Real) / ((1+Discount_Rate_Real)^Time_Period-1)

Discount_Rate_Nominal = 0.066

Discount_Rate_Real = Discount_Rate_Nominal-Inflation_Rate

Energy_Output = 1000

Inflation_Rate = 0.035

Time_Period = 30

The conditions assumed in the German low-cost scenario make PV power generation economically more attractive. At identical PV system cost levels of 15,000 DM/kWp, real cost of energy now comes down to 0.78 DM/kWh (see Table 8-8.). This level, however, is still far above the actual German retail market price.

The model yet indicates that cost-effectiveness of PV power generation can be reached at considerably higher system cost levels than in the German high-cost scenario. Here, current retail price levels can be reached at PV system costs of 4,000 DM/kWp already. Cost-effectiveness with conventional power generation options would be reached at total system costs of 1,500 to 2,000 DM/kWp, resulting in real energy costs of 0.10 and 0.08 DM/kWh, respectively.

Table 8-8. German Low-Cost Scenario Cost Analysis

Time	PV System Cost DM/kWp	Cost of Energy Nominal DM/kWh	Cost of Energy Real DM/kWh
0	15,000.00	1.16	0.78
1	14,500.00	1.12	0.75
2	14,000.00	1.08	0.72
3	13,500.00	1.04	0.70
4	13,000.00	1.01	0.67
5	12,500.00	0.97	0.65
6	12,000.00	0.93	0.62
7	11,500.00	0.89	0.59
8	11,000.00	0.85	0.57
9	10,500.00	0.81	0.54
10	10,000.00	0.77	0.52
11	9,500.00	0.74	0.49
12	9,000.00	0.70	0.47
13	8,500.00	0.66	0.44
14	8,000.00	0.62	0.41
15	7,500.00	0.58	0.39
16	7,000.00	0.54	0.36
17	6,500.00	0.50	0.34
18	6,000.00	0.46	0.31

19	5,500.00	0.43	0.28
20	5,000.00	0.39	0.26
21	4,500.00	0.35	0.23
22	4,000.00	0.31	0.21
23	3,500.00	0.27	0.18
24	3,000.00	0.23	0.16
25	2,500.00	0.19	0.13
26	2,000.00	0.15	0.10
27	1,500.00	0.12	0.08
28	1,000.00	0.08	0.05
29	500	0.04	0.03
Final	0	0.00	0.00

20.1.1.5 Conclusions on Hypothesis 1.1

After this rather elaborate assessment of photovoltaic power generation economics in a vertically integrated power market, we shall use our observations and background information derived from the previous Parts to evaluate hypothesis 1.1.

➢ Hypothesis 1.1
    In a regulated and vertically integrated market, solar electricity generation with small, grid-connected photovoltaic systems is not a cost-effective alternative for electric utilities.

Best-case and worst-case scenario models set up for Germany and the USA indicate that at current PV system price levels, photovoltaic power generation is not cost-competitive with conventional power generation technologies in vertically integrated energy markets that do not face competition. Hypothesis 1.1 can therefore be confirmed.

Our result holds true for conventional utility investment analysis based on levelized cost calculations. Competition will introduce changes into the electric utility in regard of investment decision-making parameters. In a restructured environment, utilities will have certain incentives to rethink their investment decision-making patterns. Competition will lead to low-cost strategies, initiating mergers and acquisitions. Small utilities will face customer losses and have to modify their offered services in order to stay competitive and secure revenues, profits, and a sufficient customer basis.

Mass production of photovoltaic technologies may lead to considerable cost reductions. Yet, mass production volumes could not be reached in vertically integrated markets so far. This is due to missing financial incentives to invest in new competitive technologies when protected service territories and authoritatively determined electric rates guarantee a utility's profits. Utilities investing in conventional power technologies earn sufficient revenues and do not have to fear customer losses.

In our models on PV power generation in a deregulated environment, we will examine the impacts of mass production, cost-benefit-calculation, and PV marketing strategies. It is important to note that both mass production and cost-benefit-calculation methods could have been introduced in vertically integrated power markets already. Insufficient incentives to change conventional investment decision patterns favoring conventional central power technologies undermined significant additions of photovoltaic power generation capacity. Politics, legislation, and regulatory authorities have not succeeded in setting up sufficient incentives for renewable power options. This is a crucial finding, since we will observe in the models of distributed power generation that new cost-benefit-analyses techniques correctly valuing the assets of renewable energies work best in a vertically integrated environment of electric utilities incorporating power generation, power transmission, and power distribution services.

20.1.2 Residential Customer Power Generation in a Regulated Environment

20.1.2.1 Basic Model for Residential Customer Cost Calculation

In order to calculate PV generation cost for private consumers purchasing and operating their own systems, consumers could use the levelized cost analysis introduced in the previous models. Consumers will then modify this calculation technique to include variables such as net metering or revenues through feed law legislation and full cost rates. Figure 8-13. shows the outline of our new residential PV customer model.

We assume that the net metering rate equals the utility rate without any additional charges for the electric customers. US American utilities have proposed a standby fee to eliminate customer benefits from net metering. The standby fee model, however, is rather unlikely to pass regulating authorities and legislation.[103]

---

[103] Wenger, Howard J., Thomas E. Hoff, and Jan Pepper (1996). *Photovoltaic Economics and Markets: The Sacramento Municipal Utility District as a Case Study*, p. 7-3.

Figure 8-13. Residential Customer PV Model

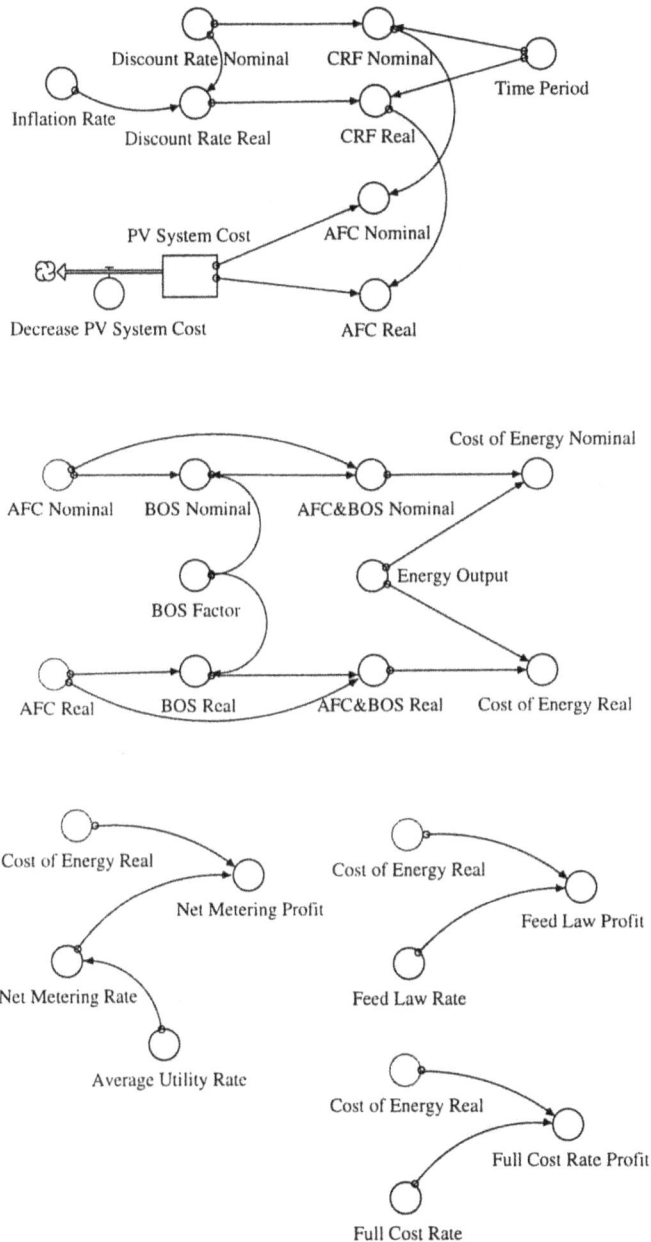

In addition to the assumptions of our previous models, the residential PV customer model sets average utility rates at 0.20 DM/kWh. Feed law legislation is expected to set the feed rate at 0.17 DM/kWh, which equals the current German feed law rate for photovoltaic power generation as presented in Part V. We also assume a full cost rate of 1.89 DM/kWh as suggested by advocates of full cost rating and implemented in various German cities.

Our model now runs on the following mathematical equations and variables:

Stock:

PV_System_Cost(t) = PV_System_Cost(t - dt) + (- Decrease_PV_System_Cost) * dt

INIT PV_System_Cost = 15000

Outflow:

Decrease_PV_System_Cost = 500

Variables and Relationships:

AFC&BOS_Nominal = BOS_Nominal+AFC_Nominal

AFC&BOS_Real = AFC_Real+BOS_Real

AFC_Nominal = PV_System_Cost*CRF_Nominal

AFC_Real = PV_System_Cost*CRF_Real

Average_Utility_Rate = 0.20

BOS_Factor = 0.015

BOS_Nominal = AFC_Nominal*BOS_Factor

BOS_Real = AFC_Real*BOS_Factor

Cost_of_Energy_Nominal = AFC&BOS_Nominal/Energy_Output

Cost_of_Energy_Real = AFC&BOS_Real/Energy_Output

CRF_Nominal =
((1+Discount_Rate_Nominal)^Time_Period*Discount_Rate_Nominal)
/ ((1+Discount_Rate_Nominal)^Time_Period-1)

CRF_Real = ((1+Discount_Rate_Real)^Time_Period*Discount_Rate_Real)
/ ((1+Discount_Rate_Real)^Time_Period-1)

Discount_Rate_Nominal = 0.08

Discount_Rate_Real = Discount_Rate_Nominal-Inflation_Rate
Energy_Output = 925
Feed_Law_Profit = Feed_Law_Rate-Cost_of_Energy_Real
Feed_Law_Rate = 0.17
Full_Cost_Rate = 1.89
Full_Cost_Rate_Profit = Full_Cost_Rate-Cost_of_Energy_Real
Inflation_Rate = 0.035
Net_Metering_Profit = Net_Metering_Rate-Cost_of_Energy_Real
Net_Metering_Rate = Average_Utility_Rate
Time_Period = 25

20.1.2.2 Sensitivity Analyses

For residential customer PV calculation, most of the variables used above are still valid. This holds true for solar radiation data and power output in kWh/kWp installed capacity; estimated inflation rate; PV system lifetime; and PV system cost. One major difference is the nominal discount rate used. Residential consumers utilize other discount rates for investment decision making than commercial electric utilities. As seen above, nominal discount rates vary for investor-owned and municipal utilities. The first has to satisfy shareholders through profit rates higher than market rates, while the latter calculates with lower cost of funds.

Usually, private consumers hold their saving capital in bank accounts at relatively low interest rates. Many consumers prefer low-risk investment options. The investment volume made by a single resident is usually lower than the capital invested by large commercial entities. For these reasons, private consumers receive lower interest rates for their investment than commercial enterprises do. This holds true as long as private customers do not invest in high-risk stocks. Our model therefore assumes that our residential customers calculate with lower nominal discount rates than commercial entities.

We run our model for nominal discount rates of 4%; 5%; 6%; 7%; and 8%, with the last value comparing to the investor-owned utility discount rate. Since the assumed inflation rate is still set at 3.5%, our lowest discount rate almost represents a zero real discount factor, that is our residential customer would be financially compensated for inflation only.

We see in Figure 8-14. that an investor-owned utility would calculate PV generation cost at real 1.11 DM/kWh, represented by line #5. Our private consumer calculates real energy cost of 1.00 DM/kWh (at 7% nominal discount rate); 0.89 DM/kWh (at 6%); 0.79 DM/kWh (5%); or 0.70 DM/kWh (4%) (lines 4 through 1).

Figure 8-14. Sensitivity Analysis on Nominal Discount Rate

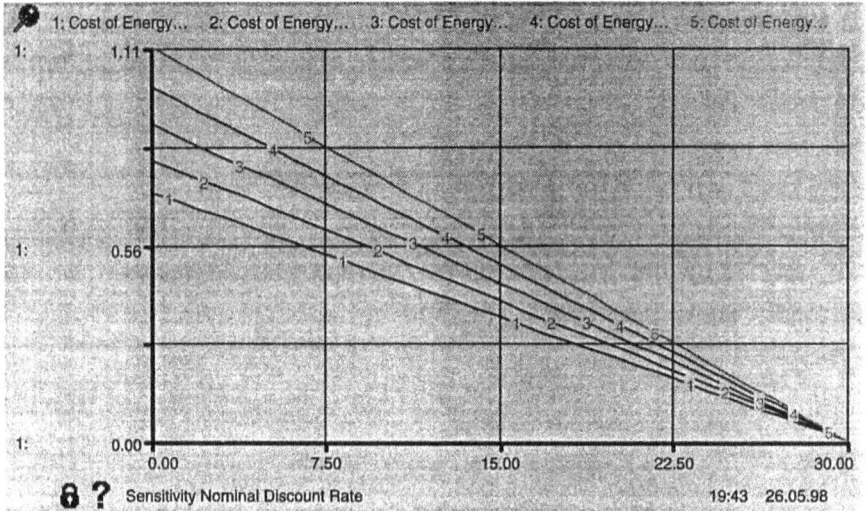

We now examine whether our residential customer loses or gains money through net metering, full cost rating, or revenues through the German feed law. Our model therefore calculates the profit to the resident through net metering by subtracting the cost of energy from the net metering rate received. Calculations for profits through full cost rates and feed law revenues are performed accordingly.

Table 8-9. illustrates the impacts of nominal discount rate variation on net metering profits to the residential customer. We see that at today's PV system cost, net metering does not compensate the residential customer for his or her investment cost. We do not reach break-even points sooner than PV system cost reductions as low as 4,250 DM/kWp at best with a nominal discount rate of 4%. Using the investor-owned utility's discount rate of 8% results in an even lower break-even PV system cost of approximately 2,500 DM/kWp.

Table 8-9. Sensitivity Analysis on Discount Rate and Net Metering Profit

		Net Metering Customer Profit in DM/kWh				
Time	PV System Cost	Nominal Discount Rate 4%	5%	6%	7%	8%
0	15,000.00	-0.50	-0.59	-0.69	-0.80	-0.91
1	14,500.00	-0.48	-0.57	-0.66	-0.77	-0.87

2	14,000.00	-0.46	-0.54	-0.63	-0.73	-0.84
3	13,500.00	-0.43	-0.51	-0.60	-0.70	-0.80
4	13,000.00	-0.41	-0.49	-0.57	-0.67	-0.76
5	12,500.00	-0.39	-0.46	-0.54	-0.63	-0.73
6	12,000.00	-0.36	-0.44	-0.51	-0.60	-0.69
7	11,500.00	-0.34	-0.41	-0.48	-0.57	-0.65
8	11,000.00	-0.31	-0.38	-0.46	-0.53	-0.61
9	10,500.00	-0.29	-0.36	-0.43	-0.50	-0.58
10	10,000.00	-0.27	-0.33	-0.40	-0.47	-0.54
11	9,500.00	-0.24	-0.30	-0.37	-0.43	-0.50
12	9,000.00	-0.22	-0.28	-0.34	-0.40	-0.47
13	8,500.00	-0.20	-0.25	-0.31	-0.37	-0.43
14	8,000.00	-0.17	-0.22	-0.28	-0.33	-0.39
15	7,500.00	-0.15	-0.20	-0.25	-0.30	-0.36
16	7,000.00	-0.13	-0.17	-0.22	-0.27	-0.32
17	6,500.00	-0.10	-0.14	-0.19	-0.23	-0.28
18	6,000.00	-0.08	-0.12	-0.16	-0.20	-0.24
19	5,500.00	-0.06	-0.09	-0.13	-0.17	-0.21
20	5,000.00	-0.03	-0.06	-0.10	-0.13	-0.17
21	4,500.00	-0.01	-0.04	-0.07	-0.10	-0.13
22	4,000.00	0.01	-0.01	-0.04	-0.07	-0.10
23	3,500.00	0.04	0.01	-0.01	-0.03	-0.06
24	3,000.00	0.06	0.04	0.02	0.00	-0.02
25	2,500.00	0.08	0.07	0.05	0.03	0.01
26	2,000.00	0.11	0.09	0.08	0.07	0.05
27	1,500.00	0.13	0.12	0.11	0.10	0.09
28	1,000.00	0.15	0.15	0.14	0.13	0.13
29	500	0.18	0.17	0.17	0.17	0.16
Final	0	0.20	0.20	0.20	0.20	0.20

Performing the same calculation for the feed law variable again indicates that PV system costs have to decline much further before the cost for photovoltaic power generation is being compensated by today's level of feed law rates. Residential customers choosing a low nominal discount rate of 4% still do not recover investment costs for PV systems more expensive than approximately 3,500 DM/kWp. For a nominal discount rate of 8%, the break-even cost is at 2,250 DM/kWp (Table 8-10.).

Table 8-10. Sensitivity Analysis on Discount Rate and Feed Law Profit

		Feed Law Customer Profit in DM/kWh				
Time	PV System Cost	Nominal Discount Rate 4%	5%	6%	7%	8%
0	15,000.00	-0.53	-0.62	-0.72	-0.83	-0.94
1	14,500.00	-0.51	-0.60	-0.69	-0.80	-0.90
2	14,000.00	-0.49	-0.57	-0.66	-0.76	-0.87
3	13,500.00	-0.46	-0.54	-0.63	-0.73	-0.83
4	13,000.00	-0.44	-0.52	-0.60	-0.70	-0.79
5	12,500.00	-0.42	-0.49	-0.57	-0.66	-0.76
6	12,000.00	-0.39	-0.47	-0.54	-0.63	-0.72
7	11,500.00	-0.37	-0.44	-0.51	-0.60	-0.68
8	11,000.00	-0.34	-0.41	-0.49	-0.56	-0.64
9	10,500.00	-0.32	-0.39	-0.46	-0.53	-0.61
10	10,000.00	-0.30	-0.36	-0.43	-0.50	-0.57
11	9,500.00	-0.27	-0.33	-0.40	-0.46	-0.53
12	9,000.00	-0.25	-0.31	-0.37	-0.43	-0.50
13	8,500.00	-0.23	-0.28	-0.34	-0.40	-0.46
14	8,000.00	-0.20	-0.25	-0.31	-0.36	-0.42
15	7,500.00	-0.18	-0.23	-0.28	-0.33	-0.39
16	7,000.00	-0.16	-0.20	-0.25	-0.30	-0.35
17	6,500.00	-0.13	-0.17	-0.22	-0.26	-0.31
18	6,000.00	-0.11	-0.15	-0.19	-0.23	-0.27
19	5,500.00	-0.09	-0.12	-0.16	-0.20	-0.24

20	5,000.00	-0.06	-0.09	-0.13	-0.16	-0.20
21	4,500.00	-0.04	-0.07	-0.10	-0.13	-0.16
22	4,000.00	-0.02	-0.04	-0.07	-0.10	-0.13
23	3,500.00	0.01	-0.02	-0.04	-0.06	-0.09
24	3,000.00	0.03	0.01	-0.01	-0.03	-0.05
25	2,500.00	0.05	0.04	0.02	0.00	-0.02
26	2,000.00	0.08	0.06	0.05	0.04	0.02
27	1,500.00	0.10	0.09	0.08	0.07	0.06
28	1,000.00	0.12	0.12	0.11	0.10	0.10
29	500	0.15	0.14	0.14	0.14	0.13
Final	0	0.17	0.17	0.17	0.17	0.17

Finally, we examine the economics of full cost rates and varying discount factors. The model results are rather surprising, with PV power generation already being economic and creating profits for the resident at today's system cost prices. Our model assumptions have led to this phenomenon. We assumed a constant full cost rate of 1.89 DM/kWp, independent from further system cost reductions. Full cost rates, however, should be set according to actual PV power generation cost level and are not designed to create profits to the PV system owner. Profit-oriented rates have been proposed as a new concept to promote investment in residential PV power generation.[104]

Table 8-11. Sensitivity Analysis on Discount Rate and Full Cost Rate Profit

| Time | PV System Cost | Full Cost Rate Customer Profit in DM/kWh | | | | |
		Nominal Discount Rate 4%	5%	6%	7%	8%
0	15,000.00	1.19	1.10	1.00	0.89	0.78
1	14,500.00	1.21	1.12	1.03	0.92	0.82
2	14,000.00	1.23	1.15	1.06	0.96	0.85
3	13,500.00	1.26	1.18	1.09	0.99	0.89

---

[104] Oberländer, Hans-Ulrich (1998). *Alternative Förderkonzepte für Photovoltaik-Anlagen*, pp. 529-533.

4	13,000.00	1.28	1.20	1.12	1.02	0.93
5	12,500.00	1.30	1.23	1.15	1.06	0.96
6	12,000.00	1.33	1.25	1.18	1.09	1.00
7	11,500.00	1.35	1.28	1.21	1.12	1.04
8	11,000.00	1.38	1.31	1.23	1.16	1.08
9	10,500.00	1.40	1.33	1.26	1.19	1.11
10	10,000.00	1.42	1.36	1.29	1.22	1.15
11	9,500.00	1.45	1.39	1.32	1.26	1.19
12	9,000.00	1.47	1.41	1.35	1.29	1.22
13	8,500.00	1.49	1.44	1.38	1.32	1.26
14	8,000.00	1.52	1.47	1.41	1.36	1.30
15	7,500.00	1.54	1.49	1.44	1.39	1.33
16	7,000.00	1.56	1.52	1.47	1.42	1.37
17	6,500.00	1.59	1.55	1.50	1.46	1.41
18	6,000.00	1.61	1.57	1.53	1.49	1.45
19	5,500.00	1.63	1.60	1.56	1.52	1.48
20	5,000.00	1.66	1.63	1.59	1.56	1.52
21	4,500.00	1.68	1.65	1.62	1.59	1.56
22	4,000.00	1.70	1.68	1.65	1.62	1.59
23	3,500.00	1.73	1.70	1.68	1.66	1.63
24	3,000.00	1.75	1.73	1.71	1.69	1.67
25	2,500.00	1.77	1.76	1.74	1.72	1.70
26	2,000.00	1.80	1.78	1.77	1.76	1.74
27	1,500.00	1.82	1.81	1.80	1.79	1.78
28	1,000.00	1.84	1.84	1.83	1.82	1.82
29	500	1.87	1.86	1.86	1.86	1.85
Final	0	1.89	1.89	1.89	1.89	1.89

The model shows one more important detail. The common value of 1.89 DM/kWh suggested by full cost rate advocates is a nominal value. The electric power industry commonly uses nominal values in order to demonstrate the non-cost-competitiveness of PV power generation. A cost calculation should, however, rather be based on real values, thus eliminating inflation rate impacts. Calculations based on real values result in considerably lower full cost rates for PV power generation.

20.1.2.3 Assessing Multiple Benefits

The Rocky Mountain Institute developed a concept of cumulated energy savings resulting in total cost savings after a threshold is reached. This "Tunneling through the Cost Barrier" suddenly completely changes the economics of energy-saving technologies.[105] In practice, the concept works as in the following example: imagine a house with a conventional oil-fired heating system. Compare to this house another one with high-efficient insulation; large thermal mass; controlled air ventilation and heat exchangers; as well as passive and active solar heating systems. While all these energy-saving features add to the cost of the second house, it may eventually reach the point where the installation of a conventional heating system is not necessary any more. If the savings from the now superfluous conventional heating system offset additional energy-saving investment, we have successfully "tunneled" through the cost-barrier.

This concept is not mere theory, but was applied in the construction of the RMI headquarters at Snowmass/Colorado. RMI's "Tunneling through the Cost Barrier" concept is based on multiple benefits resulting from one investment and can be easily applied to an economic analysis of grid-connected residential rooftop PV systems. These systems are usually being mounted on the customer's roof. The PV modules do not have to be an additional part of the house, but can be integrated into the roofing structure, thus delivering electric power *and* serving as a genuine roofing material. We have seen in Part IV that PV shingles and PV roofing tiles have been developed. Colored PV modules are commercially available as well. PV façade elements have successfully replaced conventional façade elements, while for residential housings, PV modules are usually mounted as a separate structure.

Table 8-12. exemplifies the economics of multiple benefits from multiple-purpose expenditures. Solar roofing tiles are today more expensive than conventional modules (plus 3,000 DM/kWp). Since these solar tiles replace conventional roofing material, costs for material and labor of up to 2,000 DM/kWp can be saved. At today's solar tile cost, the additional expenditures do not yet break even with the financial savings, resulting in a surplus cost of 1,000 DM/kWp.[106] These values, however, are only valid for today's very small production quantities of solar tiles.

---

[105] Rocky Mountain Institute (1997). *Tunneling Through the Cost Barrier*, pp. 1-4.

[106] Springorum, Roland, and Anne Kreuzmann (1996). *Solardachziegel. Strom aus dem Dach*, p. 13.

Costs could be significantly reduced through mass production and the use of thinfilm technologies and solar shingles.

Table 8-12. Assessing Multiple Benefits of Solar Roofing Tiles (in DM/kWp)

Additional Costs	
Additional Cost over Conventional PV Modules	3,000 DM/kWp
Saved Costs	
Saved Roofing Material	- 500 DM/kWp
Saved Mounting Material for PV Modules	- 750 DM/kWp
Saved Mounting Cost (Labor)	- 750 DM/kWp
Net Additional Cost	**1,000 DM/kWp**

Since costs vary considerably between different solar shingle and solar tile technologies, and no reliable cost-forecasts for mass-produced units are currently available, we will not incorporate this approach in our models. Furthermore, cost-savings analyses have to be performed for each individual house construction, addressing the local roof size, the roofing material, and PV components actually used.

20.1.2.4 Conclusions on Hypothesis 1.2

From our model results and the information outlined in the previous Parts, we are now able to evaluate hypothesis 1.2.

> Hypothesis 1.2
> In a regulated and vertically integrated market, solar electricity generation with small, grid-connected photovoltaic systems is not a cost-effective option for residential consumers either.

Although residential electric customers and PV system owners usually calculate with real discount rates well below utility levels, current PV system prices are yet too high to make solar electricity generation with small, grid-connected PV systems economic for residential customers. Net metering or feed law revenues do not compensate the small power producer for his or her PV engagement either. A financially attractive financing option is full cost rating. Full cost rates, however, face heavy opposition from electric utilities. Advocates for full cost rating calculate PV generation costs in nominal values, thus overestimating the real cost of PV power production, which in turn makes PV investment attractive to residential customers. Since full cost rates can be modified according to actual PV system costs, this financing system implies a certain incentive for further cost reductions. Without full cost rating, investment in small residential grid-connected PV systems is not cost-competitive with fossil fuel power generation at current PV system prices. Single customers will also not be able to effectively initiate mass production volumes and subsequent system cost reductions. Hypothesis 1.2 can therefore be confirmed, if the concept of full cost rating is not being considered.

20.2 The Economics of PV Power Generation in a Restructured and Liberalized Market Environment

20.2.1 Modeling an Electric Utility PV Market Strategy

20.2.1.1 Basic Utility Model

Our next model will examine the market situation of a comparatively small municipal utility in a deregulated power market. The small utility faces competition and declining market rates for electric power. Since the cost structure of the municipal is rather constant (the utility does not intend to merge with a larger power company to achieve economies of scale), a new strategy is necessary to guarantee the survival of the small utility. This case model may be typical for small utilities in a liberalized electric power market.

Figure 8-15. illustrates the basic outline of our municipal utility market model. The electric utility has to sell its electricity at the "Utility Rate". In a liberal retail market, electric customers compare this price level to the "Average Market Rate", that is the electric rate offered by competitors. If the utility's rate is higher than the average market rate, our utility will periodically lose electric customers ("Utility NonPV Customers").

Our municipal utility now calculates its revenues ("Utility NonPV Revenue") by multiplying average per-capita consumption ("Average Consumption") with the electric rate charged to its customers.

Figure 8-15. Basic Municipal Utility Market Model

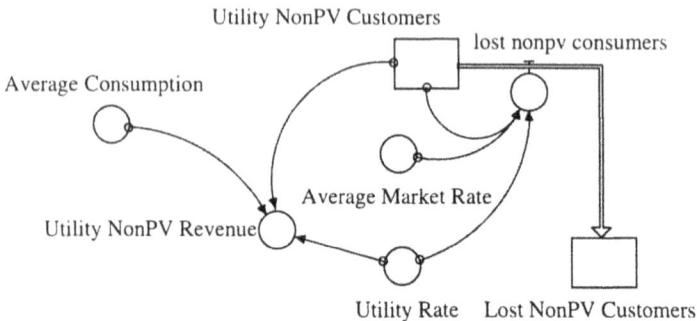

Our basic model assumes that the municipal utility serves 100,000 customers at the beginning of the deregulation process. The utility sells its power at 10 cents/kWh, while the average market rate is at 7 cents/kWh. We assume that customers compare these rates and change their electric service provider if the average market rate is lower than the utility rate. Under this condition, 10% of the utility's customers are expected to quit the municipal utility every calculation time unit (in reality, a time unit could equal a time period of one year).

Average per-capita consumption is assumed to be 1,200 kWh per time unit. This value is based on market surveys on residential electricity consumption patterns conducted by the German VDEW (Vereinigung Deutscher Elektrizitätswerke)[107].

The STELLA model then runs on the following data:

Stock 1:

Lost_NonPV_Customers(t) = Lost_NonPV_Customers(t - dt) + (lost_nonpv_consumers) * dt

INIT Lost_NonPV_Customers = 0

Inflow 1:

lost_nonpv_consumers = if Average_Market_Rate<Utility_Rate then 0.1*Utility_NonPV_Customers else 0

Stock 2:

Utility_NonPV_Customers(t) = Utility_NonPV_Customers(t - dt) + (- lost_nonpv_consumers) * dt

INIT Utility_NonPV_Customers = 100000

Outflow 2:

lost_nonpv_consumers = if Average_Market_Rate<Utility_Rate then 0.1*Utility_NonPV_Customers else 0

Variables and Relationships:

Average_Consumption = 1200

Average_Market_Rate = 0.07

---

[107] Vereinigung Deutscher Elektrizitätswerke (1998). *Drei-Personen-Haushalt Januar 1998: 91 Mark für Strom.* VDEW News, Internet Publication, http://www.strom.de/

Utility_NonPV_Revenue =
Utility_NonPV_Customers*Utility_Rate*Average_Consumption
Utility_Rate = 0.10

We run the model over 80 time units and experience the crash of the municipal's customer basis and revenues (Figure 8-16.). The number of utility customers decreases from 100,000 to some 30 customers while utility revenues collapse from US$ 12,000,000 to approximately US$ 3,636 in time period 80. In fact, our municipal utility would have been bankrupt much earlier.

Figure 8-16. Municipal Utility Customer Losses in a Deregulated Environment

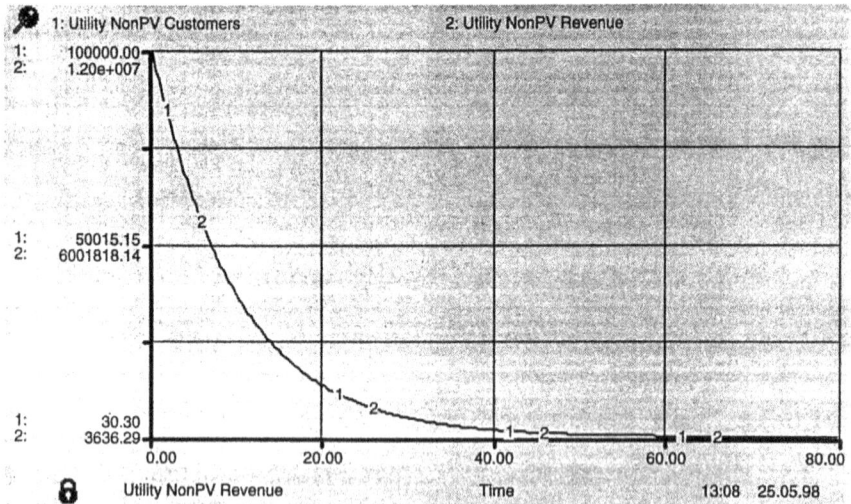

Industry experts predict a concentration process in a deregulated environment. While liberalization is usually seen as a means to introduce competition, deregulated markets tend to reduce the number of competing companies in the long run. The deregulation of the US telecommunication industry has recently led to a new concentration process with mergers and acquisitions.[108] We shall now examine how to encounter this prediction and how to avoid the municipal utility's financial bankruptcy.

---

[108] Siegele, Ludwig (1998). *Es klingelt bei den Kartellwächtern. Fusionen zwischen amerikanischen Telephongeselllschaften bedrohen den freien Wettbewerb*, p. 36.

We therefore develop a strategy to market electric power generated by photovoltaic systems. Our utility will address customers willing to pay higher than market rates for environmentally friendly power and negotiate long-term contracts. This approach is based on SMUD's PV Pioneer program.

Our model assumes that the municipal utility is able to gain a certain percentage of its electric customers as PV Pioneers. These pioneers voluntarily pay a higher than market rate for electricity generated by photovoltaic systems. Our model utility is able to sign long-term PV Pioneer contracts with 5% of its customers each calculation time unit. The PV Pioneers sign up 25-years contracts, according to the expected system lifetime. The utility still owns the systems, and the PV Pioneers only agree to pay the higher PV Pioneer electricity rate of 15 cents/kWh. The actual value of the PV Pioneer rate just influences the utility revenues, but not the model as such. The PV Pioneer rate could also be below market rates, but then the customers would stay with the utility anyway. A higher PV Pioneer rate is assumed to show that an electric utility can stay in a liberalized environment although charging higher than market prices through offering high-quality electric power (such as environmentally friendly generated electricity).

Since the contract duration limits the time of PV Pioneers staying with the municipal utility, we assume that the utility is able to regain 20% of the total lost PV Pioneers after contract termination for new contracts. In reality, the utility will acquire PV Pioneers from outside as well. To compensate for this issue, the model calculates 20% of the total lost PV Pioneers and not 20% of the lost PV Pioneers per time period. The outline of the model is illustrated in Figure 8-17.

The lower part of the model shows the development of PV capacity installations due to the PV Pioneer program. Both new PV Pioneers and regained PV Pioneers are assumed to install new PV systems. Each customer adds 1 kWp capacity. This value is based on the SMUD PV Pioneer Program that installs systems of 4 kWp on residential rooftops of California residential homes. The per-capita PV capacity addition of 1 kWp also nicely coordinates assumed average annual per-capita consumption of 1,200 kWh with PV system energy output of 1,150 kWh/kWp for New England radiation data. Our 1-kWp-per-capita installation produces approximately the amount of power consumed by the average electric customer. Our utility therefore generates the net amount of power necessary to serve its customers and does not have to import extra power[109].

---

[109] In this model case, we do not address the problem of generation load curves differing from electric consumption load curves. The calculation is just a net estimation of electric power generation and consumption during the calculation time period (here: years).

Figure 8-17. Municipal Utility PV Marketing Strategy

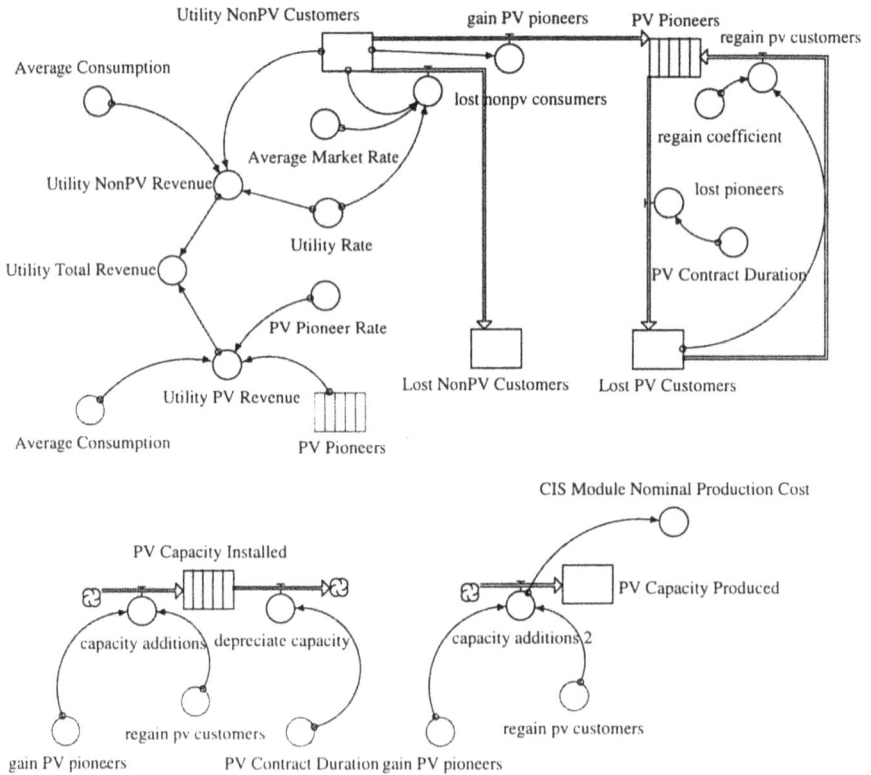

The model also calculates the reduction of PV module prices through the municipal utility PV program. We assume that the utility uses one of the latest and most cost-effective photovoltaic technologies, that is thinfilm CIS modules. It is important to remark that this calculation only indicates the module production cost, not the total system price. The latter is rather hard to predict since thinfilm modules are a new technology and economic data is hardly available. Thinfilm CIS modules differ from conventional modules in mounting and cable connecting. CIS modules can be produced in many shapes and sizes. They can be used instead of conventional roofing tiles or shingles, thus decreasing total installation costs through multiple benefits.

Table 8-13. shows a forecast of CIS module cost dependent on production volume. The data is based on studies from the Stuttgart Center for Solar Energy and Hydrogen Research and its latest PV thinfilm module generation.[110] Preparations to enter mass production are currently being negotiated.

Table 8-13. CIS Module Production Cost Forecast

Annual CIS Production Volume (kWp)	CIS Module Production Cost (US$/kWp)
25	8,500
1,200	4,000
10,000	1,500
60,000	650

With the cost data from Table 8-13., we estimate an equation for the relationship between annual CIS production volume (in kWp) and CIS module production cost (in US$/kWp) using a Microsoft Excel spreadsheet calculation. The data points can be approximated by the graph shown in Figure 8-18. and the equation

$$y = 29,146 \; x^{-0.3273}$$

with x for the annual CIS production volume and y for the estimated CIS module production cost.

Our STELLA model in this case assumes that time and calculation periods equal years, which allows us to use the CIS production cost equation that is based on annual production volumes.

---

[110] Dimmler, Bernhard, E. Gross et al. (1997). *Progress in CIGS Large Area Thin Film PV Modules Based on Industrial Process Technology,* p. 4.

Figure 8-18. PV Module Production Cost Reduction Forecast (CIS Technology)

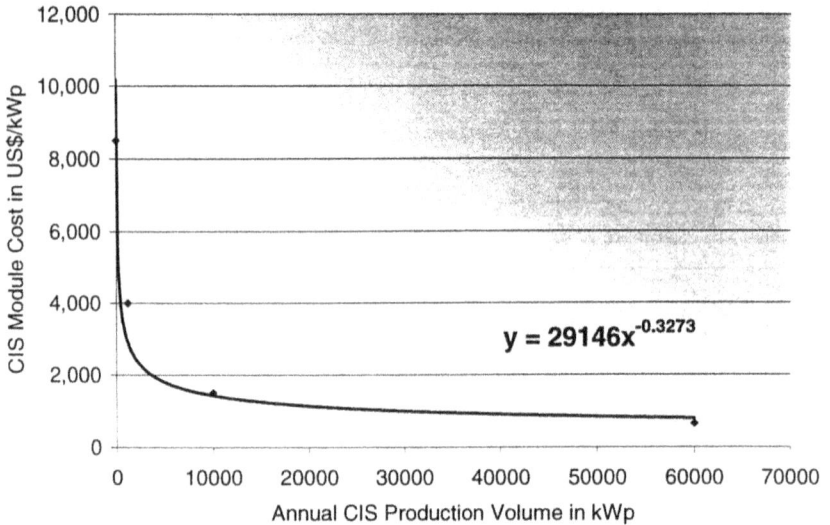

Our STELLA model now runs on the following algebraic data:

Stock 1:

Lost_NonPV_Customers(t) = Lost_NonPV_Customers(t - dt) + (lost_nonpv_consumers) * dt

INIT Lost_NonPV_Customers = 0

Inflow 1:

lost_nonpv_consumers = if Average_Market_Rate<Utility_Rate then 0.1*Utility_NonPV_Customers else 0

Stock 2:

Lost_PV_Customers(t) = Lost_PV_Customers(t - dt) + (lost_pioneers - regain_pv_customers) * dt

INIT Lost_PV_Customers = 0

Inflow 2:

lost_pioneers = CONVEYOR OUTFLOW

   TRANSIT TIME = PV_Contract_Duration

Outflow 2:

regain_pv_customers = Lost_PV_Customers*regain_coefficient

Conveyor 3:

PV_Capacity_Installed(t) = PV_Capacity_Installed(t - dt) + (capacity_additions - depreciate_capacity) * dt

INIT PV_Capacity_Installed = 0

   TRANSIT TIME = varies

   INFLOW LIMIT = INF

   CAPACITY = INF

Inflow 3:

capacity_additions = gain_PV_pioneers+regain_pv_customers

Outflow 3:

depreciate_capacity = CONVEYOR OUTFLOW

   TRANSIT TIME = PV_Contract_Duration

Stock 4:

PV_Capacity_Produced(t) = PV_Capacity_Produced(t - dt) + (capacity_additions_2) * dt

INIT PV_Capacity_Produced = 0

Inflow 4:

capacity_additions_2 = gain_PV_pioneers+regain_pv_customers

Conveyor 5:

PV_Pioneers(t) = PV_Pioneers(t - dt) + (gain_PV_pioneers + regain_pv_customers - lost_pioneers) * dt

INIT PV_Pioneers = 0

   TRANSIT TIME = varies

   INFLOW LIMIT = INF

   CAPACITY = INF

Inflows 5:

gain_PV_pioneers = Utility_NonPV_Customers*.05

regain_pv_customers = Lost_PV_Customers*regain_coefficient

Outflow 5:

lost_pioneers = CONVEYOR OUTFLOW

   TRANSIT TIME = PV_Contract_Duration

Stock 6:

Utility_NonPV_Customers(t) = Utility_NonPV_Customers(t - dt) + (- gain_PV_pioneers - lost_nonpv_consumers) * dt

INIT Utility_NonPV_Customers = 100000

Outflows 6:

gain_PV_pioneers = Utility_NonPV_Customers*.05

lost_nonpv_consumers = if Average_Market_Rate<Utility_Rate then 0.1*Utility_NonPV_Customers else 0

Variables and Relationships:

Average_Consumption = 1200

Average_Market_Rate = 0.07

CIS_Module_Nominal_Production_Cost = 29146*capacity_additions_2^(-0.3273)

PV_Contract_Duration = 25

PV_Pioneer_Rate = 0.15

regain_coefficient = 0.2

Utility_NonPV_Revenue =
Utility_NonPV_Customers*Utility_Rate*Average_Consumption
Utility_PV_Revenue = PV_Pioneers*PV_Pioneer_Rate*Average_Consumption
Utility_Rate = 0.10
Utility_Total_Revenue = Utility_NonPV_Revenue+Utility_PV_Revenue

20.2.1.2 Model Evaluation

Running our model on the assumptions above results in conventional utility customers reaching almost zero levels after 40 time periods (Figure 8-19.). We have already seen this crash in the basic municipal utility model above. However, our utility now builds up long-term customer relationships through its PV Pioneer Program. After 25 time periods (i.e. the PV Pioneer contract duration), the number of PV Pioneers peaks at 32,760 customers, approximately one third of the former electric customers to the municipal utility. This rather high level of customers can be reached although average market prices are well above utility cost levels and retail rates.

Figure 8-19. Utility Customer Class Analysis

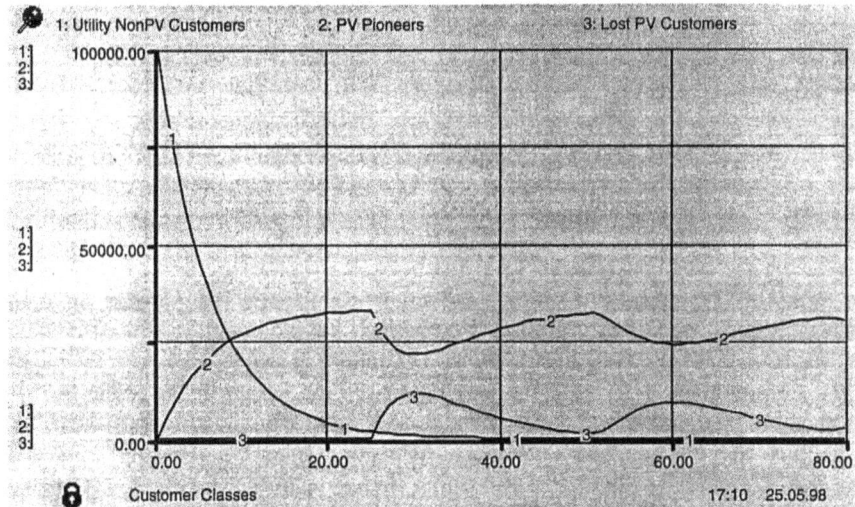

Figure 8-19. also shows that the number of PV Pioneers is oscillating after 25 time units due to contract expirations and regaining of new PV Pioneers. Running the

model for a longer time period would result in stabilizing the number of PV Pioneers to the utility.

We shall now examine the revenue impacts of the municipal utility's PV market strategy. Figure 8-20. shows how the utility's revenues from conventional power sales gradually decline, reaching zero levels at 40 time units. At the same time, utility revenues from electric sales to PV Pioneers increase, peaking at almost US$ 6,000,000 after 25 time periods. Through its PV Pioneer Program, the utility has managed to attain almost 50% of its original revenue levels. This value, again, is oscillating for a long time period and finally becoming stable.

Figure 8-20. Revenue Impact of the PV Marketing Strategy

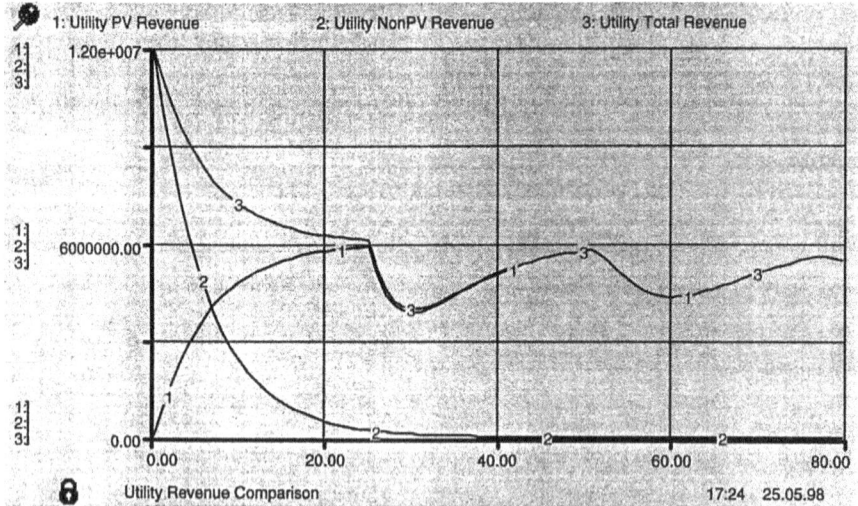

In terms of PV capacity additions, the small municipal's PV Pioneer program proves highly successful. Limited by contract duration and assuming system depreciation and removal after contract expiration, installed PV capacity peaks after 25 time units at 32,760 kWp (see Figure 8-21.). An important value for PV production cost reduction is the annual and total PV capacity produced for satisfying the utility's and customers' demands. Total PV capacity produced reaches 97,630 kWp after 80 time units, that is almost 100 MWp cumulated production volume, initiated by the PV strategy of one small electric utility. The model results add to the hopes of the investor-owned utility SMUD that a single utility is able to effectively initiate mass production through purchasing large orders of PV technologies.

Figure 8-21. Photovoltaic Capacity Impact of the PV Marketing Strategy

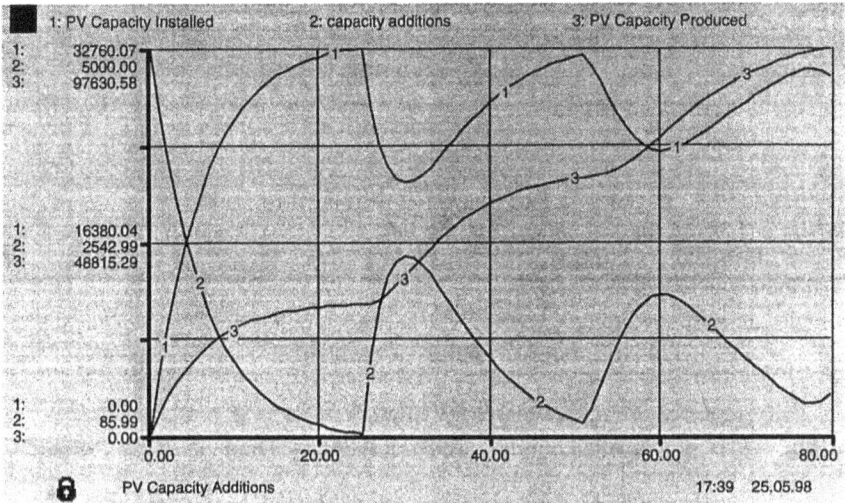

Figure 8-22. CIS Module Nominal Production Cost Reductions

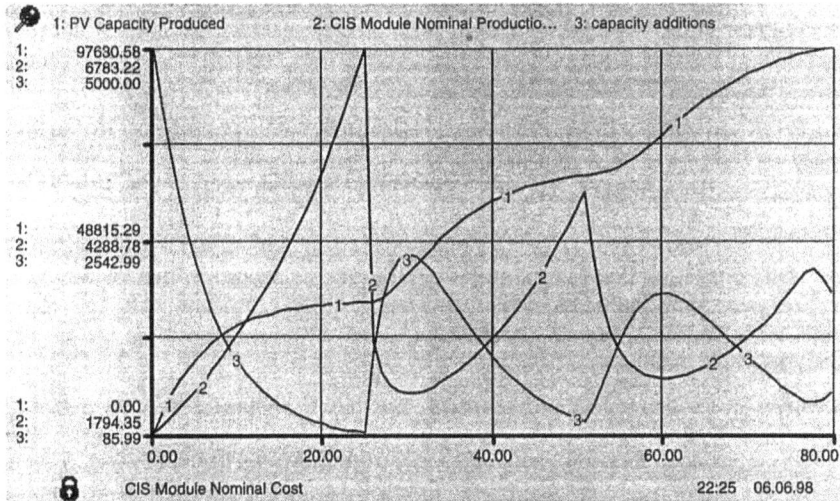

Figure 8-22. illustrates the impact of the utility's PV engagement on the development of nominal CIS module production cost. For the first 25 time periods (i.e. the timeframe when no PV Pioneer contracts have expired yet), nominal CIS

production cost increases from 1,794 $/kWp to 6,783 $/kWp. This means that our program did *not* effectively reduce system costs! We might be surprised, regarding the high amount of installed total PV capacity reaching almost 100 MWp at the end of the calculation time. The reason for this astonishing finding is that CIS module production costs are not dependent on total capacity installed, but on *annual* production volumes. As we assumed that our utility would acquire a certain percentage of its previous total customers annually, we implicitly assumed a decreasing total number of annual installations (with the number of available customers decreasing over time). To conclude, we have not yet managed to prove that our municipal's PV engagement has initiated mass production and led to lower PV system and solar electric costs.

20.2.1.3 Sensitivity Analyses

Sensitivity analyses are run to assess the impacts of varying contract duration and varying regain coefficients on the utility's total revenues. A third sensitivity analysis is to examine the impact of the utility's size, quantified by its initial number of customers when entering a deregulated power market, on its ability to actively bring down PV system and solar electric costs.

Our first sensitivity analysis varies the model variable contract duration. The value for contract duration is set at 10; 20; and 30 time units. As shown in Figure 8-23., accumulated utility revenues increase with longer contract duration. Curve #1, symbolizing a contract duration period of 10 time units, runs more smoothly, but on a lower revenue level than contract periods of 20 (curve #2) or 30 (curve #3) years.

In order to assess the impact of varying regain coefficients symbolizing the utility's ability to regain lost and gain new PV Pioneers, a sensitivity analysis is run for regain coefficient values of 0%; 10%; 20%; 30%; and 40% (curves 1 through 5,

Figure 8-24.).

The chart illustrates that the municipal utility has to regain former customers or acquire new customers in order to secure utility revenues. Higher regain coefficients result in higher accumulated total revenues, reaching up to US$ 6,000,000 each time period.

However, in the municipal utility model, maximum revenue levels do not reach higher than 50% of former utility revenues. This is due to model parameters, with PV customers regained from a limited stock of formerly lost PV customers. In reality, the municipal will try to attract new customers from outside its original service territory as well, thus being able to increase revenues theoretically without limit.

Figure 8-23. Sensitivity Analysis for Contract Duration

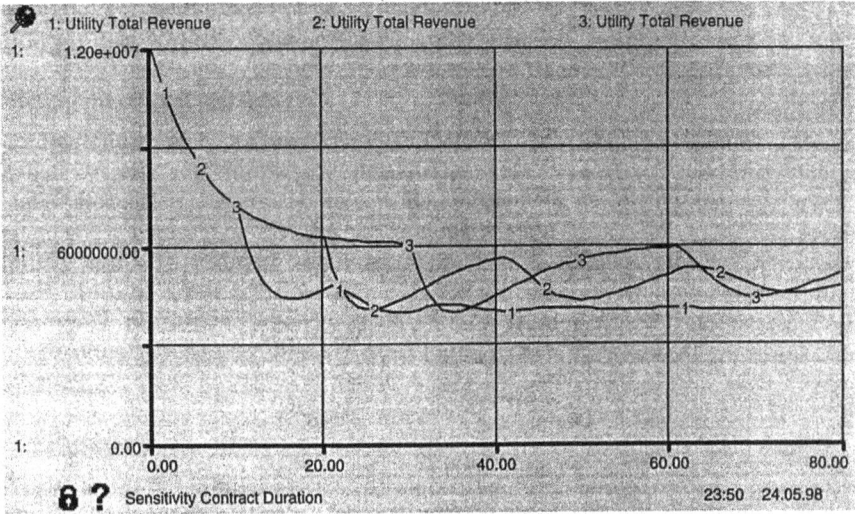

Figure 8-24. Sensitivity Analysis for Regain Coefficient

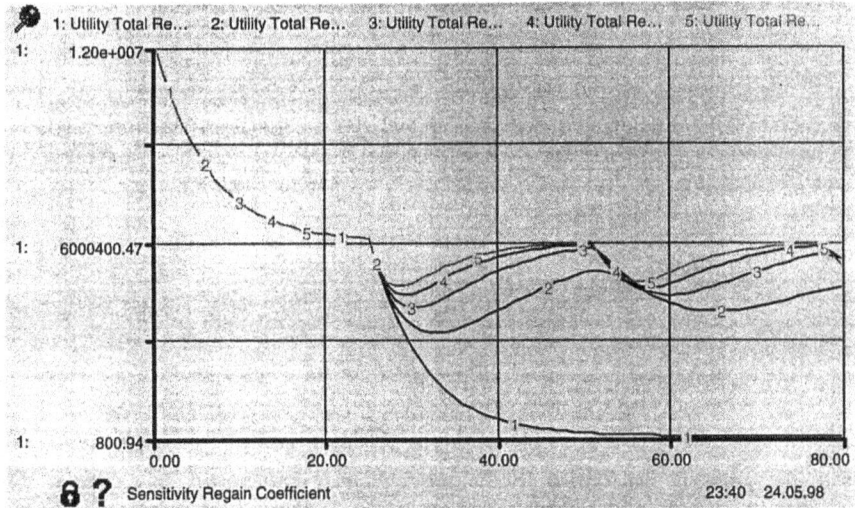

We will now perform a sensitivity analysis for utility size and PV cost reduction potential. In the previous analyses we have seen that our utility may well secure its revenues in a deregulated market due to its PV Pioneer strategy, but it was not able to initiate mass production and provide sufficient PV system orders for industrial PV component producers. We now run a new sensitivity analysis and vary the utility's initial number of electric customers.

In Figure 8-25., curves ## 1 through 5 represent initial utility customers of 100,000; 325,000; 550,000; 775,000; and 1,000,000. We observe that at the beginning, PV module costs start at values as low as approximately 1,800 $/kWp (curve #1) to 845 $/kWp (curve #5). This cost level really indicates cost-effectiveness of photovoltaic power generation. However, we are not able to maintain a sustained PV module production, with the number of our system purchases dropping each time period until reaching a long-term steady-state level due to creating a stable customer basis and renewing our PV installations after PV Pioneer contract expiration. Long-term price levels oscillate around 3,000 $/kWp (curve #1) and 1,400 $/kWp (curve #5).

Figure 8-25. Sensitivity Analysis on Initial Number of Electric Customers

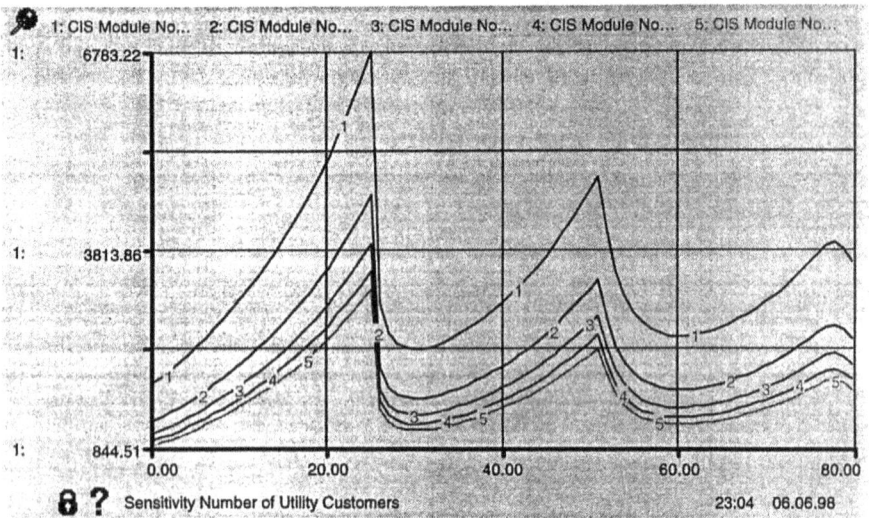

20.2.1.4 Assessing the Cost-Competitiveness of the Municipal's PV Strategy

With the analyses performed in the previous paragraphs, we have shown the impacts of a municipal utility's photovoltaic marketing strategy on various variables. In the last paragraph, we have seen that the utility's revenues are guaranteed and stabilize over time, although at a lower level than revenues in a regulated environment.

Making financial revenues, however, does not automatically mean the utility's cost-competitiveness in a liberal power market. So far, our utility's revenues from Non-PV customers were based on PV Pioneer rates set at the municipal's will. We shall now examine whether our municipal utility will stay dependent on charging higher than market rates to cover its expenses, or if it becomes cost-effective with electric costs and prices in a liberalized power market.

We will slightly modify our basic PV marketing model to examine the interrelationships between PV system installations; PV power generation cost decline through mass production; and the utility's former electric rates; average market rates; and electric rates charged from PV Pioneers.

For this model, our assumptions of the basic PV strategy model are still valid. For the calculation of the real cost of PV energy, we assume an average energy output of 1,150 kWh/kWp, as well as contract duration equaling PV system life and investment calculation time of 20 years. These assumptions are rather conservative and represent values seen with New England Electric Services above. For assessing the economics of a municipal utility, we set our nominal discount rate at 6.6%.

Our model now runs on the following algebra:

Stock 1:

Lost_NonPV_Customers(t) = Lost_NonPV_Customers(t - dt) +
(lost_nonpv_consumers) * dt

INIT Lost_NonPV_Customers = 0

Inflow 1:

lost_nonpv_consumers = if Average_Market_Rate<Utility_Rate then
0.1*Utility_NonPV_Customers else 0

Stock 2:

Lost_PV_Customers(t) = Lost_PV_Customers(t - dt) +
(lost_pioneers - regain_pv_customers) * dt

INIT Lost_PV_Customers = 0

Inflow 2:

lost_pioneers = CONVEYOR OUTFLOW

TRANSIT TIME = PV_Contract_Duration

Outflow 2:

regain_pv_customers = Lost_PV_Customers*regain_coefficient

Conveyor 3:

PV_Capacity_Installed(t) = PV_Capacity_Installed(t - dt) +
(capacity_additions - depreciate_capacity) * dt

INIT PV_Capacity_Installed = 0

TRANSIT TIME = varies

INFLOW LIMIT = INF

CAPACITY = INF

Inflow 3:

capacity_additions = gain_PV_pioneers+regain_pv_customers

Outflow 3:

depreciate_capacity = CONVEYOR OUTFLOW

TRANSIT TIME = PV_Contract_Duration

Stock 4:

PV_Capacity_Produced(t) = PV_Capacity_Produced(t - dt) +
(capacity_additions_2) * dt

INIT PV_Capacity_Produced = 0

Inflow 4:

capacity_additions_2 = gain_PV_pioneers+regain_pv_customers

Conveyor 5:

PV_Pioneers(t) = PV_Pioneers(t - dt) + (gain_PV_pioneers + regain_pv_customers
- lost_pioneers) * dt

INIT PV_Pioneers = 0

  TRANSIT TIME = varies

  INFLOW LIMIT = INF

  CAPACITY = INF

Inflows 5:

gain_PV_pioneers = Utility_NonPV_Customers*.05

regain_pv_customers = Lost_PV_Customers*regain_coefficient

Outflow 5:

lost_pioneers = CONVEYOR OUTFLOW

  TRANSIT TIME = PV_Contract_Duration

Stock 6:

Utility_NonPV_Customers(t) = Utility_NonPV_Customers(t - dt) +
(- gain_PV_pioneers - lost_nonpv_consumers) * dt

INIT Utility_NonPV_Customers = 100000

Outflows 6:

gain_PV_pioneers = Utility_NonPV_Customers*.05

lost_nonpv_consumers = if Average_Market_Rate<Utility_Rate then
0.1*Utility_NonPV_Customers else 0

Variables and Relationships:

AFC&BOS_Nominal = BOS_Nominal+AFC_Nominal

AFC&BOS_Real = AFC_Real+BOS_Real

AFC_Nominal = PV_System_Cost*CRF_Nominal

AFC_Real = PV_System_Cost*CRF_Real

Average_Consumption = 1200

Average_Market_Rate = 0.07

BOS_Factor = 0

BOS_Nominal = AFC_Nominal*BOS_Factor

BOS_Real = AFC_Real*BOS_Factor

CIS_Module_Nominal_Production_Cost = 29146*capacity_additions_2^(-0.3273)

Cost_of_Energy_Nominal = AFC&BOS_Nominal/Energy_Output

Cost_of_Energy_Real = AFC&BOS_Real/Energy_Output

CRF_Nominal =
((1+Discount_Rate_Nominal)^Time_Period*Discount_Rate_Nominal)
/ ((1+Discount_Rate_Nominal)^Time_Period-1)

CRF_Real = ((1+Discount_Rate_Real)^Time_Period*Discount_Rate_Real)
/ ((1+Discount_Rate_Real)^Time_Period-1)

Discount_Rate_Nominal = 0.066

Discount_Rate_Real = Discount_Rate_Nominal-Inflation_Rate

Energy_Output = 1150

Inflation_Rate = 0.035

PV_Contract_Duration = 20

PV_Pioneer_Rate = 0.15

PV_Share_of_Total_System_Cost = 2/3

PV_System_Cost =
CIS_Module_Nominal_Production_Cost/PV_Share_of_Total_System_Cost

regain_coefficient = 0.2

Time_Period = PV_Contract_Duration

Utility_NonPV_Revenue =
Utility_NonPV_Customers*Utility_Rate*Average_Consumption

Utility_PV_Revenue = PV_Pioneers*PV_Pioneer_Rate*Average_Consumption

Utility_Rate = 0.10

Utility_Total_Revenue = Utility_NonPV_Revenue+Utility_PV_Revenue

Figure 8-26. presents the outline of our current model. We recognize the upper part already used in our basic PV strategy model, while the lower part is derived from our former levelized cost calculation. The model now combines both approaches to assess the economics of PV power generation and PV module mass production initiated by a municipal utility's strategic engagement.

For running the STELLA model on the parameters above, our model values the share of PV module cost to equal two thirds of total system costs, that is 66%. This estimate is based on forecasts and current market values described in Part IV, Table 4-4., *PV Module Share of Total System Cost.*

Figure 8-26. PV Utility Strategy and PV Mass Production

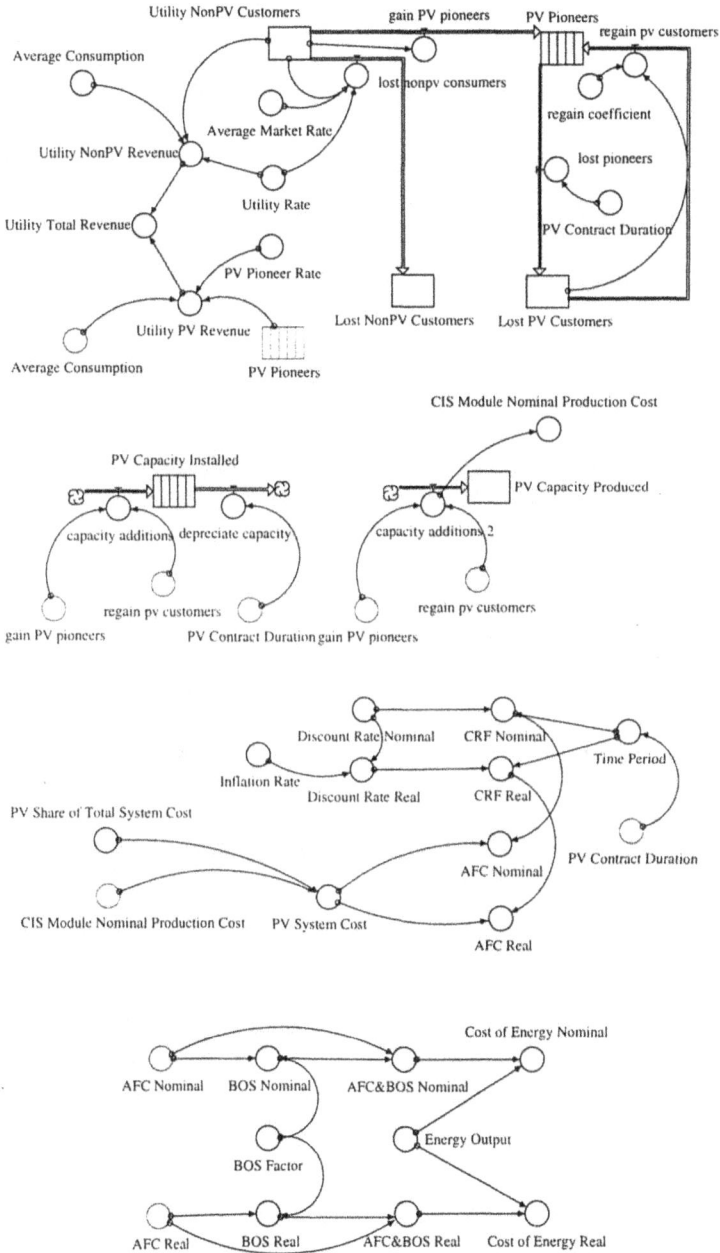

Our previous models struggled with a major problem. The number of initial customers was too low to avoid strong oscillations, which in turn impedes long-term contracts with PV manufacturers necessary for mass production. Under free market conditions, it may be feasible to acquire a rather constant number of customers each year, thus providing a stable demand for PV equipment. The necessary large volumes of PV system procurement to effectively initiate mass production and economies of scale, however, constitute a severe barrier against the cost-competitiveness of PV power generation. In the following examination, we will address this problem by extending the number of initial utility customers from 100,000 to 10,000,000. Under real circumstances, this number could also be reached by several small utilities cooperating and marketing similar PV pioneer programs.

For assessing the cost-effectiveness of the municipal utility's PV strategy, we set up three utility size categories with their number of initial customers set at 100,000; 1,000,000; and 10,000,000. In Figure 8-27., we observe that a larger number of initial electric customers substantially narrows the corridor of total system costs in subsequent calculation time periods. While a number of 100,000 customers leads to system costs up to 7,800 $/kWp, a model utility with 10,000,000 initial customers oscillates and stabilizes at approximately 1,000 $/kWp.

Figure 8-27. PV System Costs through Municipal Utility PV Strategy (in nominal $/kWp)

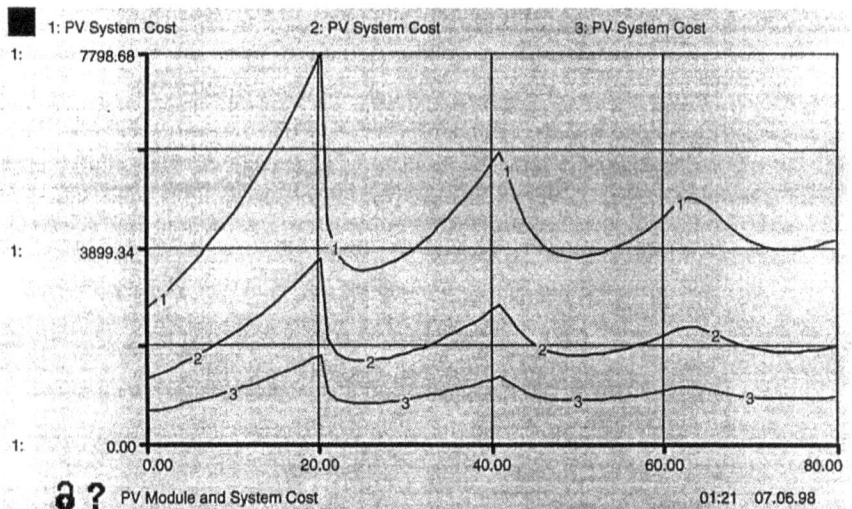

Figure 8-28. shows the case of a municipal utility with an initial stock of 100,000 electric customers. We realize that the model's PV electric generation cost (curve #1) of real 0.46 $/kWh at the curve's maximum peak and the long-term real cost level of approximately 0.25 $/kWh are far above the utility's PV pioneer rate (curve #4), i.e. the utility's revenues do not compensate PV generation costs. In this situation, the utility would have to charge a significantly higher PV Pioneer rate to cover its expenses.

Figure 8-28. Electric Rate and Cost Comparison in $/kWh
(100,000 initial customers)

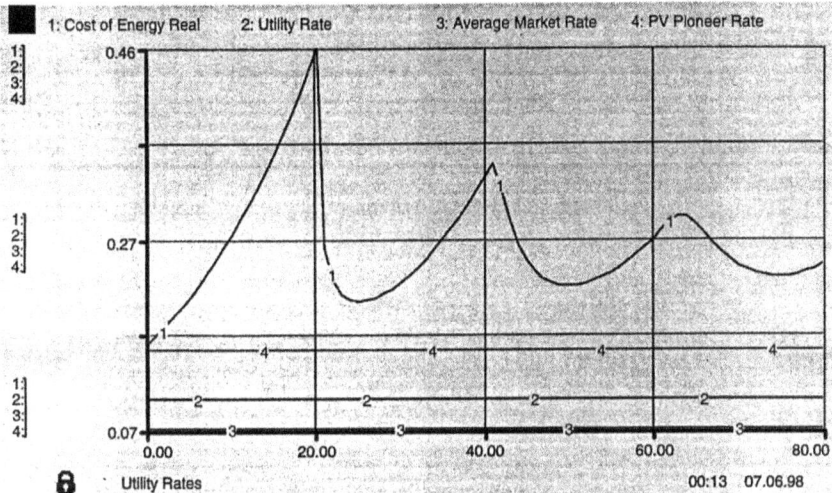

Figure 8-29. visualizes the same calculation for a customer basis of 1,000,000 consumers before electric deregulation. The oscillation corridor of PV electric generation costs is narrower than in the preceding case, peaking at 0.22 $/kWh and stabilizing at approximately 0.12 $/kWh. This cost level is between the utility's previous electric rate and its PV Pioneer rate. With some temporary exceptions, our utility is now able to cover PV generation costs through its PV Pioneer rate.

**Figure 8-29. Electric Rate and Cost Comparison in $/kWh
(1,000,000 initial customers)**

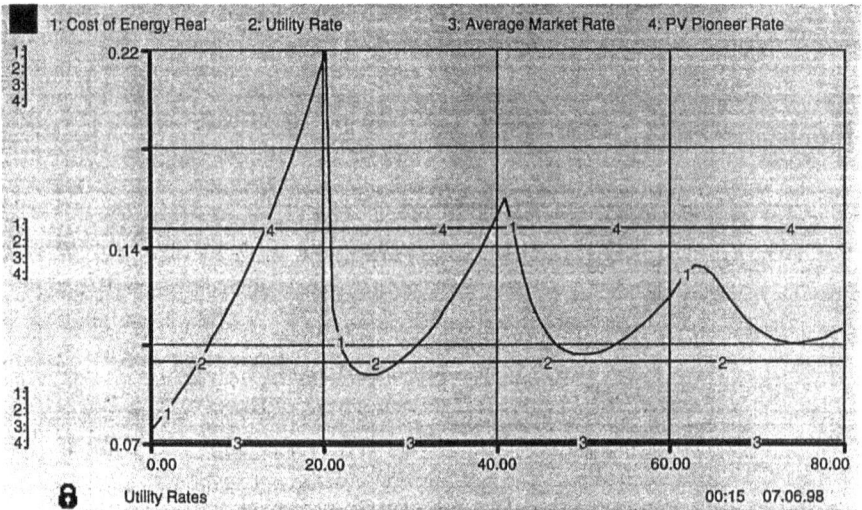

**Figure 8-30. Electric Rate and Cost Comparison in $/kWh (10,000,000 initial
customers)**

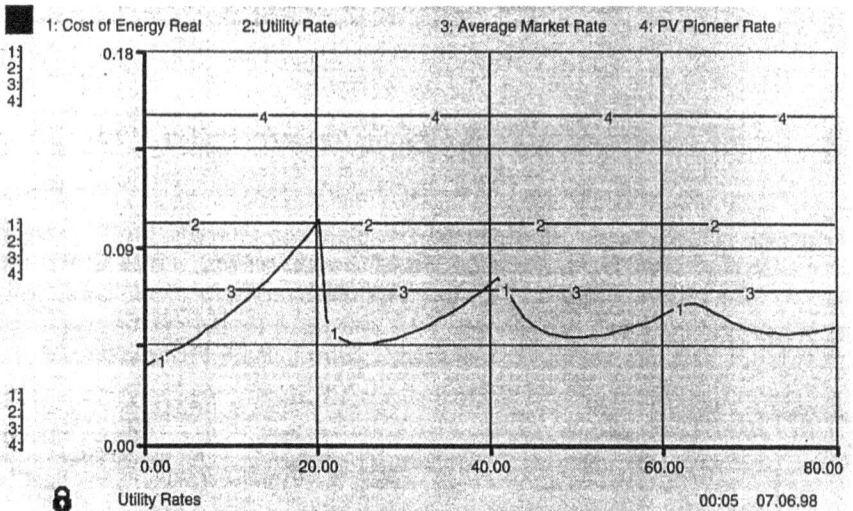

Our third example, assuming an initial customer stock of 10,000,000 persons, suggests cost-effectiveness of PV power generation in the long run. PV generation costs temporarily peak at the utility's previous cost levels of 0.10 $/kWh (curve #2, Figure 8-30.) and stabilize below projected market prices in a restructured environment (market rates set at 0.07 $/kWh). In this scenario, the utility effectively initiates mass production and cost-effectiveness of PV power generation in a liberalized environment. Oscillation margins should be as narrow as possible for guaranteeing stable long-term orders of PV systems with PV component manufacturers. Industrial module producers will not agree to invest in mass-production PV module plants without reliable procurement contracts certifying future PV module orders. As mentioned before, the necessary PV procurement volume may be hard to reach by a single small municipal utility, but could be feasible for several utilities following the same PV Pioneer strategy in order to survive in a restructured environment.

20.2.1.5 Conclusions on Hypothesis 2.1

At this point, we can evaluate hypothesis 2.1 based on our findings above.

➤ Hypothesis 2.1
   In a deregulated and liberal market, photovoltaic electricity generation will become a viable power technology for electric utilities.

For a valid evaluation, it is useful to distinguish between green pricing programs and real-cost-effectiveness approaches through mass production.

a) Revenues and profits through PV Pioneer contracts (green pricing)

In a deregulated power market, small utilities will struggle with higher costs than large utilities and face the threat of losing its electric customer basis. Our models show that small utilities can secure revenues through PV Pioneer contracts, a form of green pricing for small grid-connected PV systems at the customer's rooftops. The PV Pioneer rate, however, has to be rather high for small utilities that are not able to effectively initiate mass production of PV components. This high green pricing rate could adversely affect the acceptance of the utility's PV strategy and power contracts.

b) Cost-effectiveness with market rates through PV mass production

An electric utility can survive by embarking onto a strategic course of marketing PV power and initiating a long-term PV Pioneer program. This approach does not imply higher than market electric rates and is based on cost-effectiveness of PV power generation technologies through mass production. Two major aspects have to be regarded carefully for successfully initiating mass production and reaching cost-effective PV system price levels. First, electricity generation costs are highly dependent on local insolation levels. Unfavorable solar conditions may doom the utility's strategic solar initiative. Second, a high customer stock is necessary to reach the necessary sustained amount of annual PV capacity additions and production volumes. The number of initial electric customers is dependent on regional insolation levels and has to be larger for unfavorable solar conditions.

Under our model conditions (solar radiation data; PV Pioneer contract duration; nominal discount rate; etc.), municipal utilities need an initial customer number of 10,000,000 to initiate mass production and break-even with assumed market rates of 0.07 \$/kWh. This finding is valid for unfavorable US solar radiation conditions as for the Northeastern New England States as well as for best German radiation data (1,150 kWh/kWp electric energy output). PV generation costs will be higher for lower insolation levels (Northern Germany) and lower for areas with higher solar radiation (e.g. California). Regional differences in solar electric output therefore modify the necessary number of initial electric customers to effectively initiate mass production and reach cost-effective PV generation costs.

The necessary initial customer basis can also be reached by cooperation of several smaller municipal utilities. It is crucial to understand that large continuous capacity additions and PV volume procurements are necessary for the success of the proposed PV strategies. Regarding these findings, we can confirm hypothesis 2.1. It is yet important to note that in a deregulated and liberal power market, photovoltaic electricity generation will not automatically become a viable power technology for electric utilities. Only under certain circumstances, as indicated in our model assumptions, PV power technologies will become cost-effective.

20.2.2 Residential Customer Power Generation in a Liberal Electricity Market

20.2.2.1 Basic Model for Residential PV Power Generation

We will use the primary data from our last model case of a restructured electric environment to gain comparable results for the utility and the residential customer perspective. After all, residential customers installing and operating their own photovoltaic systems will be genuine competitors to electric utilities. In a literally deregulated power market, small power producers will have access to the public grid system. They can offer their surplus power to any commercial or residential customer and negotiate power contracts. The Independent System Operator will then grant access to the central grid through these power contracts.

Residential customer investment calculation differs from commercial utility decision making in assumptions on discount rates as discussed in the consumer model for monopolistic power markets, as well as in the single customer's impotence to initiate mass production. Single customers do not have an incentive nor the necessary investment capital to realize installation volumes necessary for economies of scale through mass production. This would be a collective task; consumer organizations have organized buying cooperatives and collective PV system orders. Greenpeace, for example, tries to market its *Cyrus* solar electric systems through collective large-scale PV system purchases.[111] The German energy consumer organization "Bund der Energieverbraucher" offers a similar program of collective system purchases and has extended its *Phönix* program from solar hot water to residential PV systems.[112]

Accepting net metering for compensating the small power producer's surplus energy fed into the grid, the residential customer's cost calculation is also dependent on the current market rate, which is supposed to decline in a deregulated environment. The residential power producer can, however, try to negotiate direct contracts with interested electric customers to sell environmentally friendly electric power at a rate above market prices. Of course, the residential power producer will face competition by our municipal electric company and its PV Pioneer strategy.

For our model, we assume that a liberal power market does no longer permit subsidizing green power generation through feed law legislation or full cost rating. Marketing green power for higher than average market rates is possible.

Our model runs on a system life time and investment calculation timeframe of 20 years. PV system costs start at 6,030 $/kWp which is the current cost level assumed by New England Electric and the Sacramento Municipal Utility District, as described above. The customer's location is to coincide with the utility's service territory, we therefore assume the same energy output of 1,150 kWh/kWp installed capacity. The inflation rate is set at 3.5%, the average market rate for electric power in a deregulated environment at 0.07 $/kWh. The electric utility that competes with the residential PV producer for customers interested in PV power offers a PV Pioneer rate of 0.15 $/kWh. Our new STELLA model now runs on the following algebraic information:

Stock 1:

PV_System_Cost(t) = PV_System_Cost(t - dt) + (- Decrease_PV_System_Cost) * dt

INIT PV_System_Cost = 6030

---

[111] Teske, Sven (1998). *Die Cyrus-Solarkampagne von Greenpeace geht in das dritte Jahr*, pp. 481-488.

[112] Kreutzmann, Anne (1997). *Phönix fliegt für Photovoltaik*, pp. 20-21.

Outflow 1:

Decrease_PV_System_Cost = 500

Variables and Relationships:

AFC&BOS_Nominal = BOS_Nominal+AFC_Nominal

AFC&BOS_Real = AFC_Real+BOS_Real

AFC_Nominal = PV_System_Cost*CRF_Nominal

AFC_Real = PV_System_Cost*CRF_Real

Average_Market_Rate = 0.07

BOS_Factor = 0.015

BOS_Nominal = AFC_Nominal*BOS_Factor

BOS_Real = AFC_Real*BOS_Factor

Cost_of_Energy_Nominal = AFC&BOS_Nominal/Energy_Output

Cost_of_Energy_Real = AFC&BOS_Real/Energy_Output

CRF_Nominal =
((1+Discount_Rate_Nominal)^Time_Period*Discount_Rate_Nominal)
/ ((1+Discount_Rate_Nominal)^Time_Period-1)

CRF_Real = ((1+Discount_Rate_Real)^Time_Period*Discount_Rate_Real)
/ ((1+Discount_Rate_Real)^Time_Period-1)

Discount_Rate_Nominal = 0.066

Discount_Rate_Real = Discount_Rate_Nominal-Inflation_Rate

Energy_Output = 1150

Inflation_Rate = 0.035

Net_Metering_Profit = Net_Metering_Rate-Cost_of_Energy_Real

Net_Metering_Rate = Average_Market_Rate

PV_Pioneer_Rate = 0.15

Residential_Advantage_over_PV_Pioneer = PV_Pioneer_Rate-
Cost_of_Energy_Real

Time_Period = 20

The model's outline is visualized in Figure 8-31.

Figure 8-31. Residential Customer Power Generation in a Liberal Electricity Market

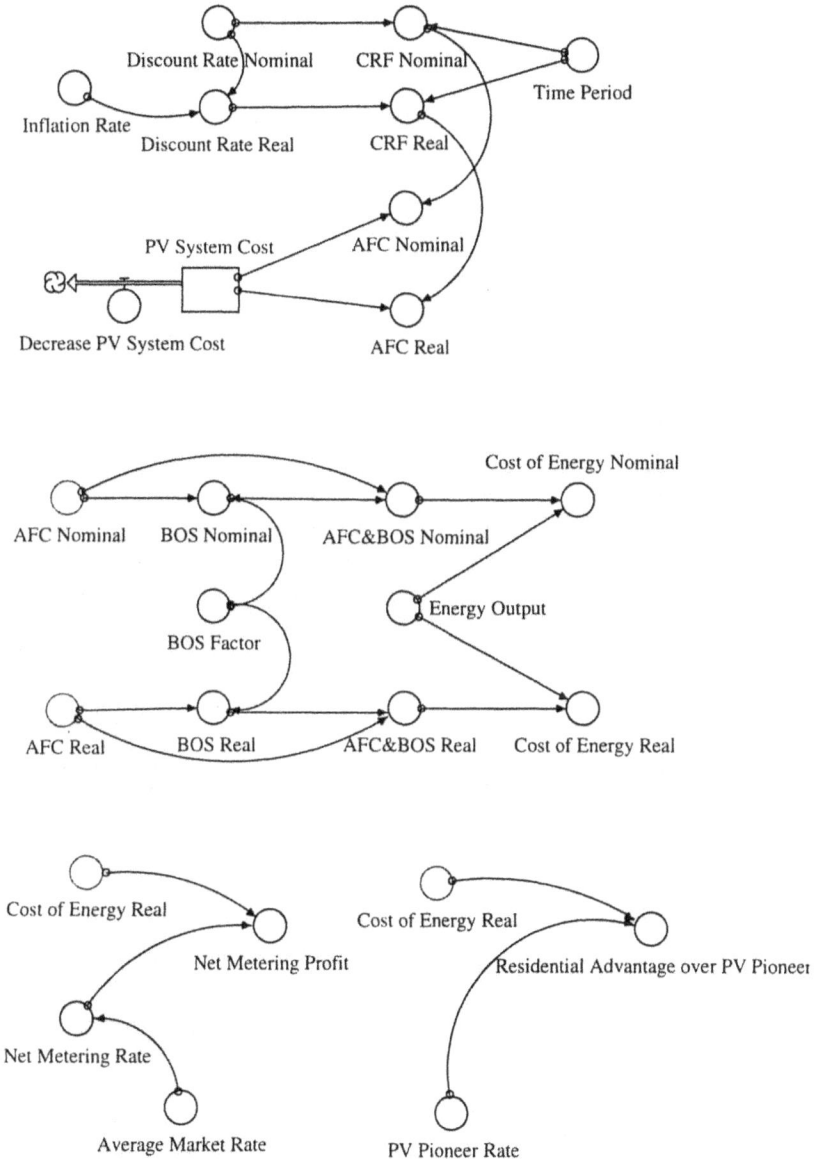

20.2.2.2 Sensitivity Analyses

To assess the real cost of residential PV power generation, we run a sensitivity analysis on nominal discount rates of 4%; 5%; 6%; 7%; and 8%. The resulting real cost of energy is shown in Figure 8-32. Line #1 represents the lowest nominal discount rate of our sensitivity run, resulting in real energy cost of 0.28 $/kWh at today's cost of 6,030 $/kWp. Total system costs are systematically decreased, reaching 3,030 $/kWp at calculation time unit 6.00 and almost zero levels at time unit 12.00. The exact values for nominal discount rate variation and real power cost are shown in Table 8-14.

Figure 8-32. Real Cost of PV Electric Power for Varying Nominal Discount Rates

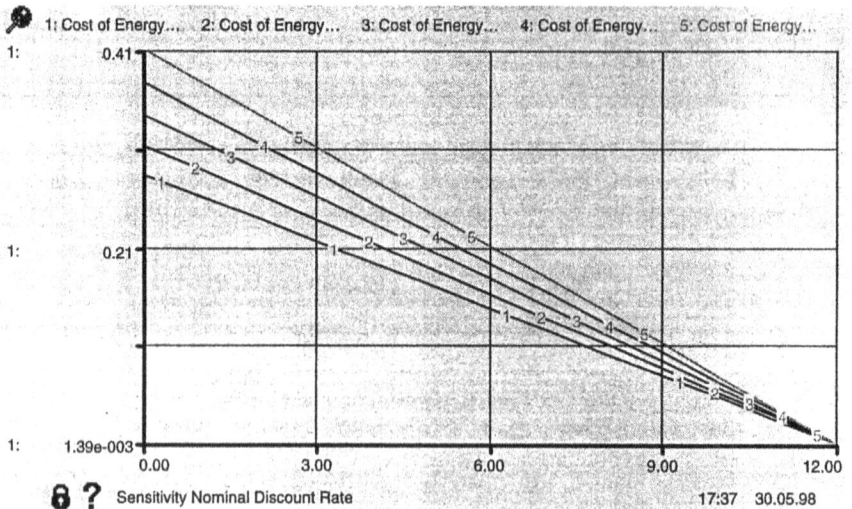

Table 8-14. Sensitivity Analysis on Discount Rate and Real Cost of Energy

| Time | PV System Cost $/kWp | Real Cost of Energy in $/kWh | | | | |
		Nominal Discount Rate 4%	5%	6%	7%	8%
0	6,030.00	0.28	0.31	0.34	0.37	0.41
1	5,530.00	0.26	0.28	0.31	0.34	0.38
2	5,030.00	0.23	0.26	0.28	0.31	0.34

3	4,530.00	0.21	0.23	0.26	0.28	0.31
4	4,030.00	0.19	0.21	0.23	0.25	0.27
5	3,530.00	0.16	0.18	0.20	0.22	0.24
6	3,030.00	0.14	0.16	0.17	0.19	0.21
7	2,530.00	0.12	0.13	0.14	0.16	0.17
8	2,030.00	0.09	0.10	0.11	0.13	0.14
9	1,530.00	0.07	0.08	0.09	0.10	0.10
10	1,030.00	0.05	0.05	0.06	0.06	0.07
11	530.00	0.02	0.03	0.03	0.03	0.04
Final	30.00	0.00	0.00	0.00	0.00	0.00

With average market prices assumed at 0.07 $/kWh, our residential power producer will not be cost-competitive until total system costs decline to at least 1,530 $/kWp levels (assuming a discount rate of 4%). The break-even system prices for all discount rates can be seen in Table 8-14. Break-even prices vary from 1,530 $/kWp to 1,030 $/kWp.

Our small residential power producer is also far from being able to recover investment costs through net metering revenues. At total system costs of 6,030 $/kWp, our resident loses from 21 to 34 cents per kWh, dependent on discount rate (Table 8-15.). In the best case (nominal discount rate 4%), total costs for photovoltaic system would have to decline to 1,530 $/kWp in order to break-even with net metering revenues.

Table 8-15. Sensitivity Analysis on Discount Rate and Residential
Net Metering Profit

		Residential Net Metering Profit in $/kWh				
Time	PV System Cost $/kWp	Nominal Discount Rate 4%	5%	6%	7%	8%
0	6,030.00	-0.21	-0.24	-0.27	-0.30	-0.34
1	5,530.00	-0.19	-0.21	-0.24	-0.27	-0.31
2	5,030.00	-0.16	-0.19	-0.21	-0.24	-0.27
3	4,530.00	-0.14	-0.16	-0.19	-0.21	-0.24
4	4,030.00	-0.12	-0.14	-0.16	-0.18	-0.20

5	3,530.00	-0.09	-0.11	-0.13	-0.15	-0.17
6	3,030.00	-0.07	-0.09	-0.10	-0.12	-0.14
7	2,530.00	-0.05	-0.06	-0.07	-0.09	-0.10
8	2,030.00	-0.02	-0.03	-0.04	-0.06	-0.07
9	1,530.00	0.00	-0.01	-0.02	-0.03	-0.03
10	1,030.00	0.02	0.02	0.01	0.01	0.00
11	530.00	0.05	0.04	0.04	0.04	0.03
Final	30.00	0.07	0.07	0.07	0.07	0.07

Our residential power producer could try to offer his electric power at a lower rate than the municipal's PV Pioneer Program that started out at the utility charging a PV Pioneer rate of 0.15 $/kWh, a level well above average market rates expected in a liberalized environment. Table 8-16. calculates the residential power producer's advantage over the utility's PV Pioneer rate. Our small power producer does not break-even until total system costs drop to as low as approximately 3,280 $/kWp in the best-case scenario with the nominal discount rate set at 4%.

Table 8-16. Sensitivity Analysis on Discount Rate and Residential Advantage over PV Pioneer Rate

		Residential Advantage over PV Pioneer Rate in $/kWh				
Time	PV System Cost $/kWp	Nominal Discount Rate 4%	5%	6%	7%	8%
0	6,030.00	-0.13	-0.16	-0.19	-0.22	-0.26
1	5,530.00	-0.11	-0.13	-0.16	-0.19	-0.23
2	5,030.00	-0.08	-0.11	-0.13	-0.16	-0.19
3	4,530.00	-0.06	-0.08	-0.11	-0.13	-0.16
4	4,030.00	-0.04	-0.06	-0.08	-0.10	-0.12
5	3,530.00	-0.01	-0.03	-0.05	-0.07	-0.09
6	3,030.00	0.01	-0.01	-0.02	-0.04	-0.06
7	2,530.00	0.03	0.02	0.01	-0.01	-0.02
8	2,030.00	0.06	0.05	0.04	0.02	0.01
9	1,530.00	0.08	0.07	0.06	0.05	0.05

10	1,030.00	0.10	0.10	0.09	0.09	0.08
11	530.00	0.13	0.12	0.12	0.12	0.11
Final	30.00	0.15	0.15	0.15	0.15	0.15

20.2.2.3 Conclusions on Hypothesis 2.2

Based on the findings in the models above, we will now evaluate hypothesis 2.2.

➤ Hypothesis 2.2
   In a deregulated and liberal market, electric utilities will face new competition from consumers producing their own photovoltaic electricity and selling it to the grid.

At today's PV system cost levels, residential customers are far from being cost-competitive with average market rates in a deregulated power market. Neither are they able to compete with a municipal utility's PV Pioneer strategy. It will therefore be hard for residential customers to effectively sell their energy through direct purchase contracts. Residential customer PV cost calculation benefits from the customer's lower than the electric utility's discount rate. However, individual small power producers willing to invest in photovoltaic technologies will hardly reach the procurement volumes necessary for cost reductions through mass production.

In a liberalized power market, marketing will be the primary instrument to sell electric power. Feed law and full cost rate revenues are rather unlikely to be implemented for being subsidy instruments initiated by either the Federal or State government or municipal authorities. Net metering is a financing concept compatible with electric restructuring legislation. However, our small residential power producer is far from being able to recover investment costs through net metering revenues at current PV system cost levels.

With cost reduction through mass production unlikely to be initiated by individual small PV power producers, hypothesis 2.2 cannot be confirmed. In a deregulated and liberal market, electric utilities will not face considerable new competition from consumers producing their own photovoltaic electricity and selling it to the grid.

20.3 The Economics of Distributed Power Generation with Photovoltaic Technologies

In the previous chapters of Part VIII, we have distinguished between vertically integrated and liberalized electric power markets. Distributed power generators that are located near the customer load side forming a rather decentralized power system can be advantageous for both vertically integrated and deregulated power markets. Technical and energetic assets occur independently from the current market situation.

It is interesting to not that the *economic* benefits of distributed power generation are easier to materialize in a vertically integrated power market where electric utilities bundle electric power generation, transmission, and distribution services. Economic benefits of distributed power generation emerge from a systemwide analysis. In a liberalized market environment, unbundling focuses an electric utility's cost calculation perspective on either generation or transmission/distribution services. Cost-savings that occur from a holistic system analysis may therefore be neglected. The same logic applies to the problems of holistic cost assessment methods such as Integrated Resource Planning (IRP), Least Cost Planning (LCP), or Demand Side Management efforts (DSM) in deregulated power markets.

On the other hand, regulated power markets have not led to an overall integration of IRP, LCP, or DSM so far. This may be due to insufficient political and legislative guidelines requiring these programs and due to the monopolistic structure of the electric power industry. Utilities did not face the threat of competition and could set their rates according to prices approved by regulatory commissions. Systemwide cost analyses would therefore have had to be performed by regulatory authorities.

Distributed power generation technologies have faced remarkable cost improvements over the last years. However, monopolistic utilities still use cost-calculation techniques developed for centralized fossil fuel power generation for their investment decision making. Cost-benefit analyses should be used to validly estimate the value of renewable power options in public grid and emerging micro-grid systems.

20.3.1 Basic Model of Distributed Power Generation

20.3.1.1 Cost-Benefit Analysis of Distributed Power Generation

Our previous models addressed the cost aspect of photovoltaic electricity generation only. Compared to conventional fossil fuel power generation technologies, PV electricity generation has advantages due to its modular and renewable character. Conventional investment calculation methods such as levelized cost calculations do not take these benefits into account.

Several benefits can be identified and have to be considered in a valid investment comparison. As discussed in Part 7, distributed electricity generation is not burdened with the cost of transmission and distribution systems. Central power units can be installed in smaller capacities when distributed generation supports a central power system.

Distributed power technologies such as photovoltaics use renewable resources. Therefore, these technologies do not need fossil fuels for electric power production and do not emit toxic gases, liquids, or other hazardous waste material.

Being located near the customer load, distributed systems minimize systemwide energy losses. For all these reasons, distributed power generation is more valuable than central power technologies. Advocates for distributed energy systems therefore favor detailed cost-benefit analyses over conventional levelized cost calculations.

The Sacramento Municipal Utility District (SMUD) identified and quantified the following benefit categories for its service territory:[113]

1. Service Revenues

In course of its Sustained Orderly Development Strategy, SMUD signed long-term contracts with PV component producers. Some of these manufacturers have opened their plants in the SMUD service area and become new electric customers to SMUD. These additional revenues would not have occurred without SMUD's strategic approach to bring down PV system costs.

The benefit category "Service Revenues" calculates net revenues to SMUD from electric power sales to the new PV manufacturing plant in its service territory.

2. REPI Federal Subsidy

Through its engagement in renewable power technologies, SMUD receives financial benefits from the US Federal Renewable Energy Production Incentive (REPI).

3. Externalities

This category quantifies the value of reduced fossil fuel emissions and related financial savings for the utility.

---

[113] Wenger, Howard J., Thomas E. Hoff, and Jan Pepper (1996). *Photovoltaic Economics and Markets: The Sacramento Municipal Utility District as a Case Study*, p. 4-1.

4. Fuel Price Risk

SMUD reduced its dependence and risk from uncertain gas price projections. Photovoltaic generators do not need fossil fuels for producing electric power.

5. Green Pricing

Each month, the participants of the PV Pioneer voluntarily pay a premium fee to the utility for having a utility-owned PV system installed on their residential rooftops.

6. Distribution and Transmission

Distributed power generators located near the customer allow to downsize the central grid or defer otherwise necessary system upgrades.

7. Generation and Energy

SMUD saves costs for central generation capacity and energy production by integrating modular photovoltaic gensets into the electricity system.

8. Losses

Locating generating units near the customer's load reduces systemwide electric losses. However, SMUD already accounted for this benefit in the previous categories. We will nevertheless add this category to our model to make it more versatile and useful for specific data of other electric utilities.

20.3.1.2 Modeling the Economics of Distributed Power Generation

Figure 8-33. illustrates the model for analyzing the economics of distributed electricity generation. This model is based on previous submodels of levelized cost calculations. This time, not total system costs (including BOS), but actual costs to the electric utility are levelized. This "Utility Cost" results from subtracting distributed benefits from total PV system costs.

The distributed benefits variable summarizes benefits from REPI; service revenues; externalities; fuel price risk mitigation; green pricing surcharges; distribution and transmission investment deferral; generation and energy cost savings; and reduced system losses.

Table 8-17. shows the monetary values of the various benefit categories for the SMUD case. SMUD has valuated each benefit category at actual utility costs (in 1996 US$). Losses are already accounted for in each benefit category, therefore our model sets this variable at 0 value.

Figure 8-33. Cost-Benefit Calculation for Distributed Power Generation

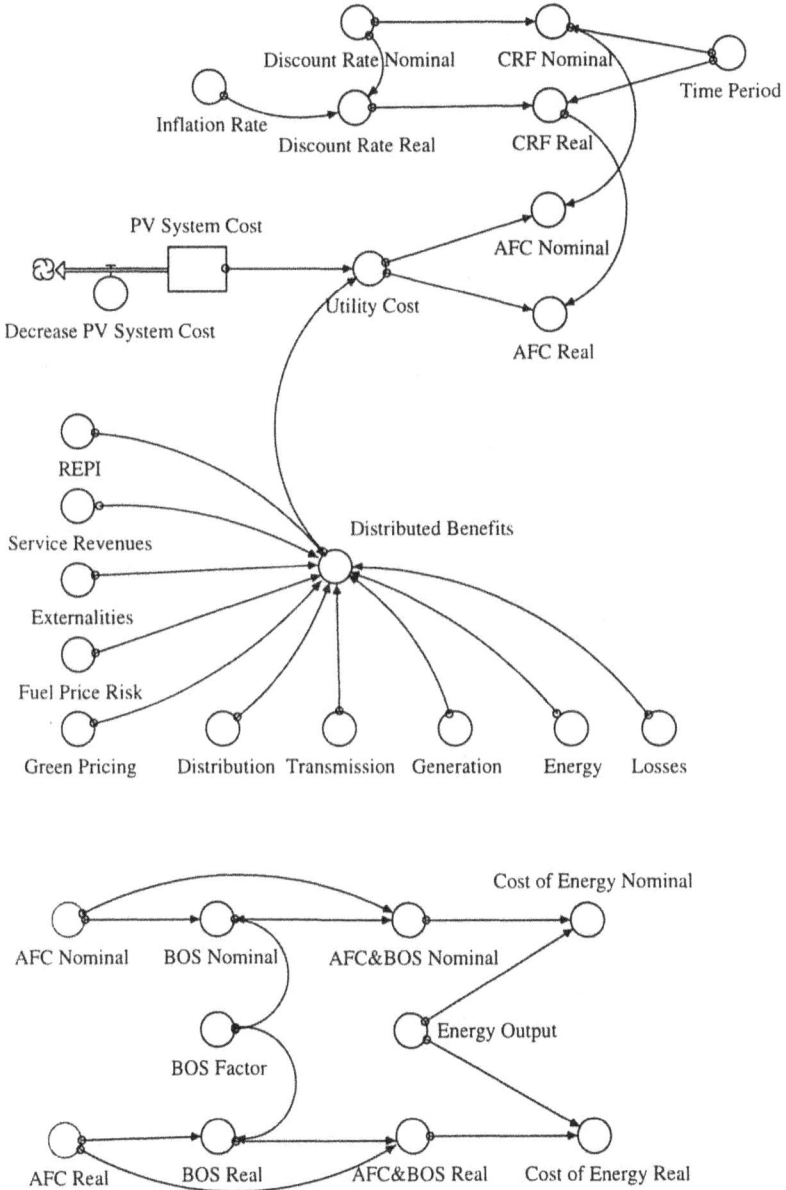

We run the model on the following variable values and relationships:

Stock:

PV_System_Cost(t) = PV_System_Cost(t - dt) + (- Decrease_PV_System_Cost) * dt

INIT PV_System_Cost = 6030

Outflow:

Decrease_PV_System_Cost = 500

Variables and Relationships:

AFC&BOS_Nominal = BOS_Nominal+AFC_Nominal

AFC&BOS_Real = AFC_Real+BOS_Real

AFC_Nominal = Utility_Cost*CRF_Nominal

AFC_Real = Utility_Cost*CRF_Real

BOS_Factor = 0

BOS_Nominal = AFC_Nominal*BOS_Factor

BOS_Real = AFC_Real*BOS_Factor

Cost_of_Energy_Nominal = AFC&BOS_Nominal/Energy_Output

Cost_of_Energy_Real = AFC&BOS_Real/Energy_Output

CRF_Nominal =
((1+Discount_Rate_Nominal)^Time_Period*Discount_Rate_Nominal)
/ ((1+Discount_Rate_Nominal)^Time_Period-1)

CRF_Real = ((1+Discount_Rate_Real)^Time_Period*Discount_Rate_Real)
/ ((1+Discount_Rate_Real)^Time_Period-1)

Discount_Rate_Nominal = 0.066

Discount_Rate_Real = Discount_Rate_Nominal-Inflation_Rate

Distributed_Benefits =
SUM(Distribution,Energy,Externalities,Fuel_Price_Risk,Generation,
Green_Pricing,Losses,REPI,Service_Revenues,Transmission)

Distribution = 117

Energy = 805

Energy_Output = 1810

Externalities = 340

Fuel_Price_Risk = 201

Generation = 315

Green_Pricing = 44

Inflation_Rate = 0.035

Losses = 0

REPI = 221

Service_Revenues = 708

Time_Period = 30

Transmission = 55

Utility_Cost = PV_System_Cost-Distributed_Benefits

Table 8-17. Benefit Category Description and Valuation for SMUD's Grid-Connected Residential PV Systems

Benefit Category	Description	SMUD Valuation ($/kW, 1996)
Service Revenues	Net service revenues from a new local PV manufacturing plant	708
REPI	Federal Renewable Energy Production Incentive	221
Externalities	Value of reduced fossil fuel emissions	340
Fuel Price Risk	Value of reduced risk from uncertain gas price projections	201
Green Pricing	Monthly contributions from PV Pioneers	44
Distribution	Deferral of distribution capacity investment	117
Transmission	Deferral of transmission capacity investment	55
Generation	Avoided marginal cost of systemwide generation capacity	315
Energy	Avoided marginal cost of systemwide energy production	805

Losses	Electric loss reduction	included in categories above
Total		**2,806**

20.3.1.3 Sensitivity Analysis Results

A sensitivity analysis on PV system costs reveals highly interesting results. Even at today's PV system costs of more than 6,000 $/kWp, real cost of energy for distributed PV power generation is as low as 0.09 $/kWh. Respecting the value of distributed power options to SMUD, this means that real PV power cost reach today's retail price level in California. Cost-effectiveness with conventional power generation options is reached at system cost levels of 4,000 to 5,000 $/kWp, resulting in real energy costs of 0.03 $/kWh and 0.06 $/kWh, respectively (Table 8-18.).

Running the sensitivity analysis for PV system cost reductions in 500 $/kWp steps offers one more surprise. Nominal and real cost of energy become *negative* at present value system costs of approximately 2,800 $/kWp, i.e. SMUD's estimated PV benefit value. For system costs above this break-even point, the cost of energy implicitly sets the minimum retail rate that the electric utility has to charge for its electricity (at no extra profit margin except for the profit incorporated in the chosen discount rate). At cost levels below the break-even point, the utility is not dependent on payments from residential customers for the amount of power produced at distributed sites. Investment in grid-integrated PV technologies pays for itself through its benefit value. PV capacity installations at critical points of the utility grid help to reduce total utility system costs that emerge from central power generation.

This finding, of course, does not mean that PV capacity additions would be economic at any scale. The higher the share of distributed PV capacity becomes, the lower the benefits from adding more PV systems, finally reaching zero or even negative values. Distributed benefits have to be quantified according to the share of distributed power capacity in the utility's overall energy system.

Furthermore, this calculation is valid from a utility-specific perspective only. Other utilities may experience considerably lower distributed benefits than SMUD and therefore reach the break-even point at lower PV system cost levels. The model however indicates the cost-effectiveness for distributed power technologies to SMUD in the utility's service territory. Distributed energy options such as photovoltaics can have economic advantages in complex centralized electric power systems. These benefits result from deferring capacity additions and from environmentally friendly power generation at lower risks than fossil fuel alternatives and have to be accounted for to improve utility investment decision-making processes.

Table 8-18. Distributed Cost-Benefit Calculation

Time	PV System Cost $/kWp	Distributed Benefits $/kWp	Utility Cost $/kWp	Cost of Energy Nominal $/kWh	Cost of Energy Real $/kWh
0	6,030.00	2,806.00	3,224.00	0.14	0.09
1	5,530.00	2,806.00	2,724.00	0.12	0.08
2	5,030.00	2,806.00	2,224.00	0.10	0.06
3	4,530.00	2,806.00	1,724.00	0.07	0.05
4	4,030.00	2,806.00	1,224.00	0.05	0.03
5	3,530.00	2,806.00	724.00	0.03	0.02
6	3,030.00	2,806.00	224.00	0.01	0.01
7	2,530.00	2,806.00	-276.00	-0.01	-0.01
8	2,030.00	2,806.00	-776.00	-0.03	-0.02
9	1,530.00	2,806.00	-1,276.00	-0.05	-0.04
10	1,030.00	2,806.00	-1,776.00	-0.08	-0.05
11	530.00	2,806.00	-2,276.00	-0.10	-0.06
12	30.00	2,806.00	-2,776.00	-0.12	-0.08
Final	0.00	2,806.00	-2,806.00	-0.12	-0.08

In Table 8-18., one interesting observation should be given credit to. Our model calculated the utility's break-even point at approximately 2,800 $/kWp. This value nicely coincides with the US Utility PhotoVoltaic Group's assumption that PV system cost levels of 3,000 $/kWp would launch a $27 billion market in the USA.[114] Our model shows that this value is based on real cost-benefit utility analyses and has to be taken seriously.

20.3.1.4 Conclusions on Hypothesis 3.1

Having identified and modeled the benefits of distributed power generation, we are now able to evaluate hypothesis 3.1.

---

[114] Vesey, Andrew (1997). *2,000 by 2,000: A Challenge*, p. 2.

> Hypothesis 3.1
Electric restructuring and technological advancement will for economic reasons create a rather decentralized (distributed) electric power system.

In our models and in the SMUD case study, we have seen that distributed power generation has economic advantages compared to central power systems. Distributed generation with photovoltaic technologies located near the customer load can be cost-effective due to several financial benefits that have to be accounted for. Specific calculations are necessary for each utility, since benefit categories and values are mostly case-specific. Revenues from green pricing vary between programs; central grid congestion and constraints are site-specific; and benefits from service revenues depend on the utility's commitment to a sustained PV power strategy.

Other benefit categories may be universally applicable, such as fuel price risk mitigation and benefits from distribution and transmission upgrade deferral as well as line losses.

In a restructured electricity market, unbundling could decrease the economic benefits from distributed power generation from a single utility's perspective, since distributed benefits mainly occur from an integrated systemwide perspective (transmission, distribution, etc. benefits). It is interesting to note that cost-benefit analyses more effectively contribute to the use of distributed power technologies in vertically integrated than in liberalized and unbundled power markets. Some benefits, however, occur for authentic power producers, e.g. service revenues; Federal and State subsidies; green pricing premiums; externality benefits (reduced fossil fuel emissions and hence lower taxes); risk mitigation aspects; electric loss reduction). Distributed power production with photovoltaic technologies therefore has economic benefits in a deregulated environment as well, although on a lower scale than in vertically integrated power markets.

Our findings partly confirm hypothesis 3.1. State-of-the-art distributed photovoltaic technologies that are implemented near the customer load have economic benefits and can modify our centralized power systems toward a more distributed power system. Electric restructuring, however, can be seen as a liability rather than an asset to promote decentralized power generation.

20.3.2 Model for Micro-Grid Economics

20.3.2.1 Micro-Grid Components

Micro-grids are a special case of distributed power generation. Micro-grids break with the premise of large-scale public grid systems. An economic analysis has to consider a micro-grid's specific characteristics.

Various technologies can be used to form a micro-grid. Our model will assess the economics of residential photovoltaic systems and micro-grid compatible technologies as suggested by Hoff, Wenger, Herig, and Shaw.[115] The authors propose micro-grid components such as energy-efficiency (energy-efficient lights; AC tune-up; energy-efficient refrigerators; gas dryers) and natural gas powered fuel cells next to photovoltaic systems.

Table 8-19. lists the suggested components of a micro-grid and their electricity cost in $/kWh. As for the electricity generated by photovoltaic modules, we will use our previous model calculations. The table indicates that investment in energy-efficient appliances is a crucial prerequisite for setting up micro-grid systems. We also observe that the electric cost of fuel cell power generation is highly dependent on their operating mode. Fuel cells in cogeneration mode (i.e. systems producing heat and electricity at the same time) achieve electricity costs of 0.081 $/kWh, while costs for fuel cells operating in non-cogeneration mode reach as high as 0.163 $/kWh.

Table 8-19. Cost Comparison of Micro-Grid Components

	Lights	AC Tune-Up	Fuel Cell (cogen.)	Refrige-rator	Gas Dryer	Fuel Cell (non-cogen)
Capital Cost ($)	200	250	2,000	750	500	N/A
O&M Cost (c/kWh)			1.0			1.0
Grid Cost (c/kWh)			1.5			1.5
Fuel Cost (c/kWh)			N/A		3.0	9.6
Size (kWp)			2.0			N/A
Life (a)	7	10	11	10	10	7
Elec. Production/ Savings (kWh/a)	750	600	4,400	1,200	1,200	2,350
Electricity Cost ($/kWh)	0.050	0.057	0.081	0.085	0.087	0.163

---

[115] Hoff, Thomas E., Howard J. Wenger, Christy Herig, and Robert W. Shaw (1998). *A Micro-Grid With PV, Fuel Cells, and Energy Efficiency*, pp. 1-6.

It is important to note that data for fuel cell electricity generation is based on estimates assuming the availability of 2-kWp residential cogeneration fuel cells on the market. Up to date, no such systems are commercially available.

20.3.2.2 Modeling Micro-Grid Economics

To derive a model for assessing the economics of micro-grid technologies, we again modify our basic model on levelized cost calculation.

Since micro-grids are by definition not connected to the public utility system, we do not have benefits from distributed power generation as discussed in our previous model any more. We are now merely interested in the real cost of PV power generation and the specific cost of other micro-grid components in order to decide whether PV is a cost-effective micro-grid technology.

We construct a model assuming current PV system costs of 6,030 $/kWp; PV system energy output at 1,810 kWh/kWp; nominal discount rate 6.6% at an inflation rate of 3.5%; as well as system life coinciding with the calculation's time frame of 30 years.

As for the electricity costs of micro-grid components, we refer to the data in Table 8-19.

Our STELLA model now runs on the following variables and relationships:

Stock 1:

PV_System_Cost(t) = PV_System_Cost(t - dt) + (- Decrease_PV_System_Cost) * dt

INIT PV_System_Cost = 6030

Outflow 1:

Decrease_PV_System_Cost = 500

Variables and Relationships:

AC_Tune_Up = 0.057

AFC&BOS_Nominal = BOS_Nominal+AFC_Nominal

AFC&BOS_Real = AFC_Real+BOS_Real

AFC_Nominal = PV_System_Cost*CRF_Nominal

AFC_Real = PV_System_Cost*CRF_Real

BOS_Factor = 0

BOS_Nominal = AFC_Nominal*BOS_Factor

BOS_Real = AFC_Real*BOS_Factor

Cost_of_Energy_Nominal = AFC&BOS_Nominal/Energy_Output

Cost_of_Energy_Real = AFC&BOS_Real/Energy_Output

CRF_Nominal =
((1+Discount_Rate_Nominal)^Time_Period*Discount_Rate_Nominal)
/ ((1+Discount_Rate_Nominal)^Time_Period-1)

CRF_Real = ((1+Discount_Rate_Real)^Time_Period*Discount_Rate_Real)
/ ((1+Discount_Rate_Real)^Time_Period-1)

Discount_Rate_Nominal = 0.066

Discount_Rate_Real = Discount_Rate_Nominal-Inflation_Rate

Energy_Output = 1810

Fuel_Cell_Cogen = 0.081

Fuel_Cell_NonCogen = 0.163

Gas_Dryer = 0.087

Inflation_Rate = 0.035

Lights = 0.050

Refrigerator = 0.085

Time_Period = 30

Figure 8-34. illustrates the STELLA model's graphical outline.

Figure 8-34. Electricity Costs of Micro-Grid Components

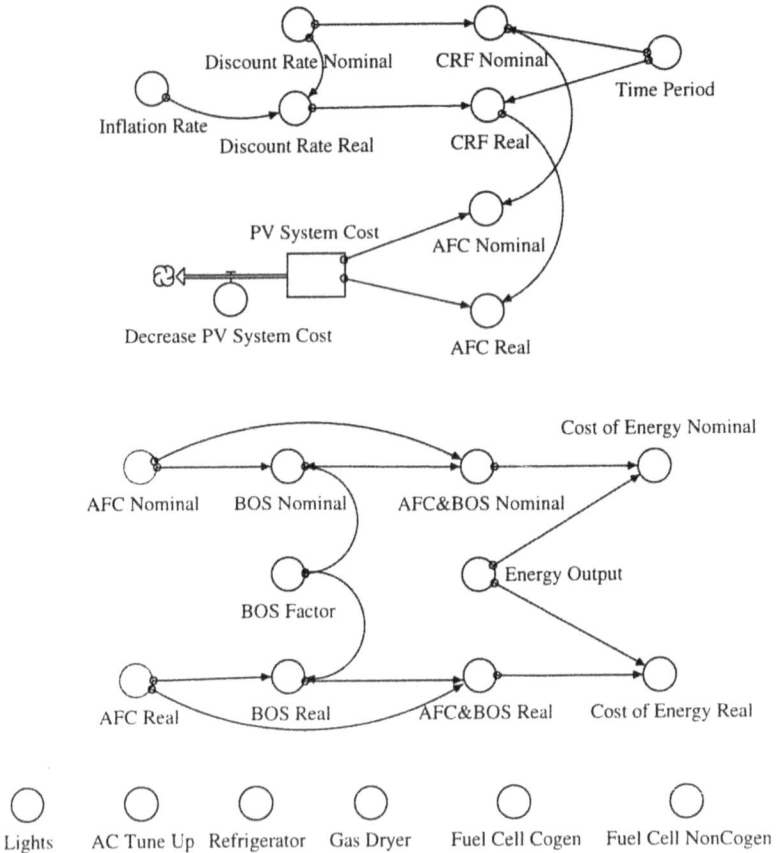

20.3.2.3 Model Analysis

The micro-grid components can be categorized into two groups. Energy-efficiency efforts try to cut down the load side of the micro-grid, while fuel cells and PV systems provide the necessary electricity to power appliances. These two categories are substitutes only up to a certain degree. Decreasing the energy consumption in the micro-grid system may become uneconomic from a certain point. On the other hand, energy-efficiency efforts are more cost-effective in the first place compared to installing expensive power generation capacity incrementally. We therefore examine energy-efficiency efforts and electric power generation technologies separately.

Figure 8-35. presents the model's results comparing electricity cost for PV power generation and energy efficiency efforts. Installing energy-efficient gas-dryers and refrigerators costs 0.087 $/kWh and 0.085 $/kWh, respectively. PV system costs have to reach approximately 3,000 $/kWp to economically break-even with these two micro-grid components. In other words, for PV system costs higher than 3,000 $/kWp, it is more cost-effective to invest in energy-efficient refrigerators and gas dryers before sizing and installing the PV system.

Energy-efficient lights and AC tune-up cost as low as 0.050 $/kWh and 0.057 $/kWh, respectively. In our model, PV system costs would have to decrease to approximately 2,000 $/kWp in order to break-even with these micro-grid components.

Our observations once more prove the basic rule used by PV system installers to first reduce the load side of a residential home as low as possible before sizing the PV system according to the remaining energy demand. Energy-efficiency investments are generally more cost-effective than installing large PV systems to provide power to energy-inefficient appliances.

Figure 8-35. Electricity Cost for PV Power Generation and Energy Efficiency Efforts

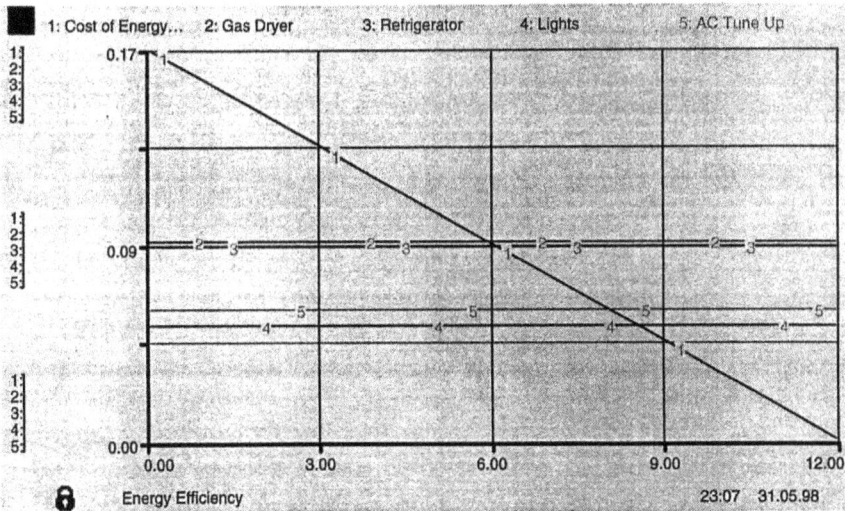

However, the remaining energy-demand has to be satisfied by electric power generators. Cutting down energy-demand to zero levels would neither technically nor economically be advisable in micro-grid systems. Figure 8-36. illustrates the economics of PV power generation in comparison to natural gas fuel cell power production in cogeneration mode and in non-cogeneration mode. We see that in our model, PV systems are not far from being cost-effective with non-cogeneration fuel cells, breaking even at approximately 5,800 $/kWp installed photovoltaic capacity. Fuel cells operating in cogeneration mode, however, are a more cost-effective option. To break-even with this technology, PV system cost would have to reach 2,800 $/kWp levels, producing electric power at 8 cents/kWh.

Concise values for our model calculations can be derived from Table 8-20.

Figure 8-36. Electricity Cost for PV and Fuel Cell Power Generation

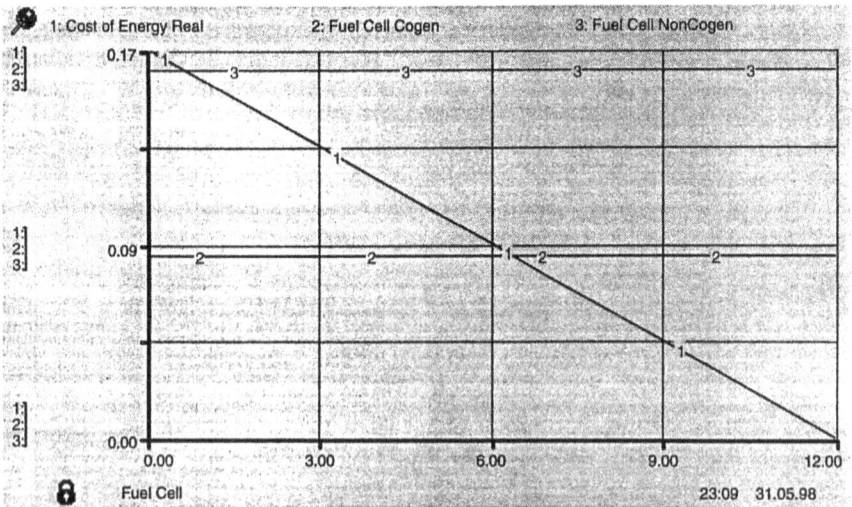

Table 8-20. Electric Costs of Micro-Grid Components (in $/kWh)

Time	PV System Cost $/kWp	Real Cost of PV Energy	Gas Dryer	Refrige- rator	Lights	AC Tune- Up	Fuel Cell Cogen	Fuel Cell NonCogen
0	6,030.00	0.172	0.087	0.085	0.050	0.057	0.081	0.163
1	5,530.00	0.158	0.087	0.085	0.050	0.057	0.081	0.163
2	5,030.00	0.144	0.087	0.085	0.050	0.057	0.081	0.163

3	4,530.00	0.129	0.087	0.085	0.050	0.057	0.081	0.163
4	4,030.00	0.115	0.087	0.085	0.050	0.057	0.081	0.163
5	3,530.00	0.101	0.087	0.085	0.050	0.057	0.081	0.163
6	3,030.00	0.087	0.087	0.085	0.050	0.057	0.081	0.163
7	2,530.00	0.072	0.087	0.085	0.050	0.057	0.081	0.163
8	2,030.00	0.058	0.087	0.085	0.050	0.057	0.081	0.163
9	1,530.00	0.044	0.087	0.085	0.050	0.057	0.081	0.163
10	1,030.00	0.029	0.087	0.085	0.050	0.057	0.081	0.163
11	530	0.015	0.087	0.085	0.050	0.057	0.081	0.163
Final	30	0.001	0.087	0.085	0.050	0.057	0.081	0.163

In our model, we did not assess the economics of conventional small power generators such as diesel gensets commonly used in off-grid applications. We assumed that fuel cells would be a more promising and future-oriented technology with higher relevancy and lower possible cost due to current research and development efforts in the automotive industry. Fuel cells also have another advantage over conventional gensets: they can be operated with hydrogen, therefore being independent from volatile fossil fuel prices and taxation issues. Hydrogen technologies are most important for micro-grid applications since hydrogen can be generated by photovoltaic electrolysis, hence providing a storage medium for solar generated electric power.

Technicians like to regard photovoltaics and fuel cell technologies as integral parts of micro-grid systems serving as supplements. This is due to their individual peaking time, with photovoltaic energy production peaking in summer and fuel cell cogeneration providing highest electric output in wintertime when comparatively more heat for space heating has to be produced.

20.3.2.4 Conclusions on Hypothesis 3.2

We are now ready to evaluate our last hypothesis 3.2. We therefore refer to the models and data of Chapter 18.3, which includes both distributed power generation and the special case of micro-grid systems.

➢ Hypothesis 3.2
   Small, grid-connected photovoltaic systems on residential rooftops will be a cost-effective part of distributed power systems.

In the models of distributed power generation, we have seen that grid-connected photovoltaic technologies can contribute cost-effectively to a more distributed power system due to systemwide economic benefits. Although site-specific data and regional differences have to be considered in an economic cost-benefit analysis, empiric data from the Sacramento Municipal Utility District and our theoretical models confirmed the cost-effectiveness of PV technologies for selected applications even at today's PV system costs.

Electrically isolated micro-grids are a special case of distributed power generation. Our analysis struggles with missing data on necessary technologies. Small hydrogen fuel cells are an integral part of micro-grids, but are not yet commercially available in the necessary size and capacity. Under our model conditions, PV system costs would have to drop to 2,000 $/kWp to financially break-even with other micro-grid components. However, since photovoltaic technologies, fuel cells, and energy efficiency efforts can only be substituted to a certain degree, a mere cost-comparison is not suitable to determine the economic feasibility of PV technologies for micro-grid systems.

With micro-grid systems being but a special case of distributed power systems, our findings confirm hypothesis 3.2. In certain applications, small, grid-connected photovoltaic systems on residential rooftops can be a cost-effective part of distributed power systems. Cost-benefit analyses indicate that residential rooftop PV systems are cost-effective at selected utility grid locations even at today's PV system prices.

PART IX.

RESULTS AND OUTLOOK

21. Résumé of Hypotheses Evaluation

21.1 PV Economics in Vertically Integrated and Regulated Electricity Markets

We have seen in best-case and worst-case scenario models set up for Germany and the United States that at current PV system price levels, photovoltaic power generation is not cost-competitive with conventional power generation technologies in vertically integrated energy markets that do not face competition. Our result holds true for conventional utility investment analysis based on levelized cost calculations and the absence of mass production cost levels. Insufficient financial and structural incentives have so far favored electric utility investment in conventional central rather than in distributed photovoltaic power generation.

Current PV system prices are also too high to make solar electricity generation with small, grid-connected PV systems economic for residential customers. Lower discount rates as well as net metering or feed law revenues do not compensate small residential power producers for their PV engagement. A financially attractive option is full cost rating.

21.2 PV Economics in Restructured and Liberalized Electricity Markets

Photovoltaic power generation may benefit from the new market situation in a restructured and liberalized environment. Small utilities struggling with higher than market costs can secure revenues and acquire a reliable customer basis through PV Pioneer contracts. With this green pricing approach, the PV Pioneer rate, however, has to be rather high for small utilities that are not able to effectively initiate mass production of PV components.

A strategy of actively bringing down PV system costs through mass production can circumvent the need for higher than market rates compensating for the utility's PV engagement. However, both sufficient solar radiation and a large initial customer stock are necessary to effectively initiate mass production and secure sustained PV module procurement volumes. The necessary initial customer basis can also be reached by cooperation of several smaller municipal utilities.

In a deregulated and liberal market, electric utilities will not face considerable new competition from consumers producing their own photovoltaic electricity and selling it to the grid. At today's PV system cost levels, residential customers are far from being cost-competitive with average market rates in a deregulated market environment. Feed law and full cost rate revenues are rather unlikely to be implemented in a liberalized power market. Net metering is a viable financing concept compatible with electric restructuring legislation, but does not sufficiently compensate residential photovoltaic power producers for their investment.

21.3 PV Economics in Distributed Electricity Systems

In our models and in empiric examples, we have seen that distributed power generation has economic advantages compared to central power systems. Distributed generation with photovoltaic technologies located near the customer load can be cost-effective even at today's PV cost levels due to various financial benefits that have to be accounted for.

In a restructured electricity market, unbundling could decrease the economic benefits from distributed power generation from a single utility's perspective, since some major distributed benefits occur from an integrated systemwide perspective only.

Micro-grids are a feasible power option to areas without a central grid system. To make electrically isolated micro-grids cost-competitive with existing central systems, further cost-reductions of PV technologies and commercialization of residential-size fuel cells and hydrogen technologies are needed.

22. Economic Prospects and Problems for Photovoltaic Power Generation

22.1 System Cost Reductions through Mass Production

In course of this research work and doctoral thesis, we have seen that today's cost for photovoltaic technologies is yet too high to make grid-connected PV power generation cost-effective with conventional fossil fuel alternatives. Except for the models on cost-benefit analyses, significant cost reductions are necessary to make residential grid-connected PV systems a cost-competitive power generation option in both regulated and restructured energy markets.

New technologies have been developed that promise considerable cost-reduction potential. Thinfilm solar cells require smaller quantities of semiconductor material and can be used for various PV module technologies such as conventional large-area PV modules; solar shingles; solar roofing tiles; and façade elements.

One of the most promising and comparatively non-toxic cell materials is thinfilm copper indium diselenide (CIS). Large production quantities could make PV power generation cost-effective with central fossil fuel technologies. However, large and sustained annual procurement volumes are essential to effectively initiate mass production and lead to considerable cost reductions. Depending on location and market conditions for solar power generation, annual PV production volumes of 60 MWp would be necessary to reach break-even cost levels of conventional power generation. This PV capacity would result in module production costs as low as 650 US$/kWp according to the Stuttgart Research Center for Hydrogen and Solar Technologies (ZSW). It has to be noted that 60 MWp of *annual* PV capacity installations are required, not cumulated PV shipments.

22.2 New Utility Investment Calculation Methods

While most of this work's findings suggest that mass production has yet to decrease PV system costs before PV power generation will be a cost-competitive investment option for both utilities and residential power producers, small distributed PV systems already are cost-effective at selected locations of the electric power grid. Conventional levelized cost calculations do not consider the *benefits* of modular PV power units for the total utility grid system. The Sacramento Municipal Utility District estimates its benefits from residential grid-connected PV systems as high as 2,806 US$/kWp. This reduces real net electric generation costs for PV power in the SMUD service area to 0.09 US$/kWh and makes SMUD's PV systems cost-effective with current California electric retail rates (at 0.10 US$/kWh) at today's PV technology cost.

However, this finding should not be generalized. Benefit categories and values are highly dependent on local solar radiation levels; specific electric utility parameters (discount rate; constraints in utility transmission and distribution capacity; strategic PV engagement; etc.); subsidy and legislative issues (Federal or State renewable energy incentives; upcoming restrictive environmental legislation; etc.). While some benefit categories may be universally applicable (systemwide loss savings; fuel price risk mitigation), a case-by-case assessment has to carefully evaluate utility-specific benefit categories and values.

It is interesting to observe that cost-benefit analyses work best in a regulated environment of vertically integrated utilities. Several benefit categories only materialize into economic values from a utility's business perspective when electric generation, transmission, and distribution services are integrated in one entity. Deregulation and unbundling of electric services may consequently diminish the financial benefits of PV power generation for electric power producers that do not own and operate transmission and distribution systems. Benefit categories such as distribution and transmission upgrade deferral are cost-relevant from an integrated systemwide perspective only.

22.3 Electric Utility PV Strategy in a Deregulated Environment

Cost-benefit assessment tools may effectively increase the number of PV power installations, but lose much of their stimulus in a deregulated and unbundled electric power system.

In the US, we currently experience significant changes in the structure of the energy sector toward competition and unbundling of power services. The legislation of the European Union also urges European countries to open their markets and liberalize national electricity systems. It is therefore advisable to examine the role of solar electricity in liberalized power markets and develop strategies for renewable power production.

In 1998, Germany has started to liberalize its electricity industry. While large utilities have prepared the coming changes with mergers and acquisitions on a national and international scale, small power producers and municipal utilities will face vital problems in the starting lowest-cost-competition. Small utilities will hardly be able to financially compete with large power producers and therefore gradually lose their customer basis that was previously protected in their service territory.

We have developed several models to simulate customer losses through competition and assessed a strategy to guarantee a small utility's survival in terms of revenues; profits; number of customers; and long-term service contracts. Embarking on a sustained photovoltaic marketing strategy, our small utility successfully creates long-term customer relationships. Large numbers of participants in the utility's photovoltaic program are necessary to initiate and sustain mass production of PV power technologies. If the utility succeeds in this strategy to actively bring down PV system costs, it may reach PV power generation cost levels that are necessary to compete with expected market prices. If the utility cannot successfully achieve the necessary mass-production volumes, it may try to charge higher than market rates through marketing green power.

Based on our models' assumptions, an initial utility basis of 10,000,000 electric customers is necessary to reach the required annual PV procurement volumes and make PV power generation cost-competitive in a liberalized power market. Under real circumstances, this number could also be reached by several small utilities cooperating and marketing similar PV pioneer programs.

The success of the utility's PV engagement depends on its strategic approach. A strategic cooperation between the electric utility, residential customers, and PV system producers is necessary for making PV technologies an economically feasible power generation option. In this context, it may be illustrative to quote Hans-Jörg Bullinger, director of the Fraunhofer Institute for Labor and Organizational Research, Stuttgart, Germany:

"In the future, cooperations will be characterizing our industrialized world: cooperations between enterprises and their customers; their industrial suppliers; and even their competitors."[116]

While on a large scale, this development would mean a serious threat to economic competition, the basic idea can be helpful to promote environmentally friendly technologies that would not reach their marketability otherwise. A utility - PV

---

[116] Bullinger, Hans-Jörg (1998). Mega-Konzerne bald mächtiger als Staaten, p. 5.

producer cooperation in terms of long-term reliable PV system procurements is necessary to guarantee sustained reliable PV system purchases and functions as an incentive and stimulus for the PV component manufacturer to set up new PV module production plants. On the other hand, long-term contracts between electric utilities and electric customers are the basis for installing PV systems at the customers' rooftops. Using today's technology, these systems work for several decades, which should be represented by the contract duration.

It has been argued that vertical cooperations between utilities, PV installers, and PV manufacturers, could not effectively reduce PV system costs, but would lead to a crowding-out of small firms and to concentration in the industry.[117] This argument may hold true for cooperations between utilities and PV system installers. In Germany, we currently observe electric utilities extending their service programs and electric craftsmen demonstrating against the utilities' engagement in their service area.

In the context of our examination, cooperations between utilities and PV manufacturers are a more relevant issue. In fact, utility - PV manufacturer cooperations and long-term purchase agreements would crowd-out competitors that are not able to acquire these long-term procurement contracts, thus having no guarantee of being able to sell their products in the future. There is no real argument against this concern. Furthermore, large volumes of one single technology (e.g. CIS thinfilm) are necessary for effectively reaching mass production volumes. Today, the industry conducts research and development on a broad variety of semiconductor materials. Strategic cooperations could lead to a preselection of a specific technology and reduce the number of competing companies.

Some empiric evidence can help to mitigate these concerns. The Sacramento Municipal Utility District's SOD strategy is based on long-term contracts with several manufacturers and introduces competition for annual procurement volumes through a bidding system.[118] This enhances competition between PV component manufacturers. Furthermore, several manufacturers could invest in parallel into the most promising technologies. While this approach does not avoid a concentration and reduction of semiconductor materials, it does allow competition between a large quantity of PV component manufacturers.

Although these concerns have to be regarded carefully, cooperations seem to be one of the very few options to successfully initiate mass production of PV power technologies in a deregulated electricity market.

---

[117] Seltmann, Thomas (1995). *Brauchen wir Einkaufsgemeinschaften?*, p. 24.

[118] Wenger, Howard J., Thomas E. Hoff, and Jan Pepper (1996). *Photovoltaic Economics and Markets: The Sacramento Municipal Utility District as a Case Study*, p. vii.

23. Outlook

23.1 The Scale Issue of Economic Activity

The energy policy of industrialized countries focuses on securing accessibility of low-price fossil fuel resources. Both military and economic combats have been conducted to reach this targets, from national as well as multinational perspectives.

Each country has a certain environmental capacity. Up to these capacity limits, the regional biological and ecological systems are able to compensate for natural resource consumption and emissions from economic activities. Industrialized countries largely depend on importing fossil fuels. Prof. Herman E. Daly, University of Maryland, criticizes the boundless import of natural carrying capacity.

> "Trade between nations or regions offers a way to loosen local constraints by importing environmental services (including waste absorption) from elsewhere. Within limits, this can be quite reasonable and justifiable. But when carried to extremes in the name of free trade it becomes destructive. It leads to a situation in which each country is trying to live beyond its own absorptive and regenerative capacities by importing these capacities from elsewhere. ... How it would be possible for all countries to be net exporters of goods and net importers of carrying capacity is not explained... A subsystem cannot grow beyond the scale of the total system of which it is a part. ... The scale of the economy must remain below the capacity of the ecosystem sustainably to supply services such as photosynthesis, pollination, purification of air and water, maintenance of climate, filtering of excessive ultraviolet radiation, recycling of wastes, etc."
>
> Herman E. Daly (1996), Beyond Growth, pp. 165-166.

Regarding actual problems of heavy regional environmental pollution and the threat of global climate change, we have reached the point where the size of our global economy has reached the limits of the environment's carrying capacity. While our conventional market economies effectively allocated resources and social legislation tries to compensate for unequal distribution of wealth, we have not yet solved the *scale* problem that haunts our national and global economies.

Redirecting our economy toward decentralized systems such as distributed power generation; distributed water supply; and distributed resource systems; could help to use limited natural resources more considerately, respecting local assets and constraints. To a certain degree, imports and exports of natural resources may still be advisable. Using local resources, however, reduces systemwide energetic losses and helps to keep the scale of economic systems within the boundaries of the ecological carrying capacity. Decentralized energy systems are a vital component of sustained economic activities.

23.2 The Entropy Law and Solar Energy

Nicholas Georgescu-Roegen proposed a new economic view by integrating physical laws into economic concepts. The laws of thermodynamics serve as the theoretical framework to rethink mainstream economic theories.

Georgescu-Roegen argues that the economic system has to be regarded as a subsystem of a larger physical system including our planet and the sun. With his entropy hourglass, the author visually applies the laws of thermodynamics to economic activity.[119] In a closed system represented by the hourglass, there is neither creation nor destruction of sand (see Figure 9-1.). The first law of thermodynamics addresses this phenomenon as conservation of matter/energy.

Figure 9-1. Entropy Hourglass by Georgescu-Roegen

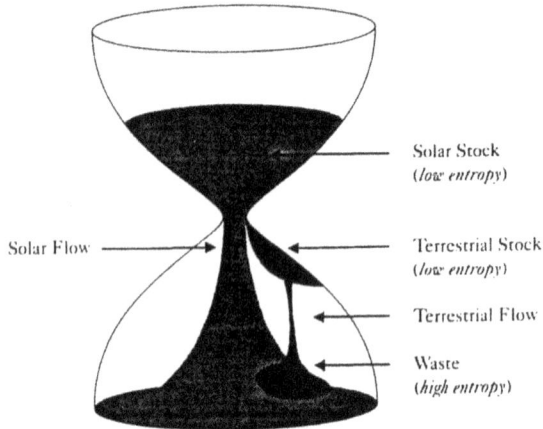

Solar Stock
(*low entropy*)

Solar Flow

Terrestrial Stock
(*low entropy*)

Terrestrial Flow

Waste
(*high entropy*)

Daly, Herman E. (1996). *Beyond Growth.*
The Economics of Sustainable Development, p. 29.

---

[119] Daly, Herman E. (1996). *Beyond Growth. The Economics of Sustainable Development*, pp. 27-30.

Over time and use, the entropy of both solar and terrestrial resources increases in a closed system. This is the analog of the second law of thermodynamics; entropy increases in an isolated system. The quality of matter/energy degrades until the available resources become unusable for any further purposes. From a microeconomic perspective, transformation processes and industrial activity may add to the value of the goods produced. However, the same activities represent but a destruction of useful low-entropy matter/energy from a thermodynamic perspective.

The thermodynamic analysis indicates two major findings applicable for this research. First, fossil fuels are limited as any form of terrestrial resources. Using fossil fuels for electric power generation increases systemwide entropy and makes these resources unavailable for usage in future applications. Since we do not yet know whether fossil fuels will be used for purposes of higher importance than generating heat (for which they can be easily replaced by other energy sources, dependent on economic valuation), we should be considerate depleting fossil fuels at today's extraction rate.

The second finding refers to the availability and use of solar energy. Solar radiation is the only resource that is not limited from the terrestrial perspective. The hourglass tries to show the relationship between the amount of energy available from the sun and the matter/energy incorporated in the terrestrial stock. Georgescu-Roegen argues that "given the disproportion between the amount of energy available from the sun compared to that in the earth, the industrial phase of man's evolution will cease long before the sun stops shining. The higher the degree of economic development, the sooner the end will come."[120]

While this is a rather skeptical long-term perspective, we should recognize that the only resource that is not as limited as most terrestrial resources is solar energy. Increasing the share of renewable energies (which are in principle all based on solar energy) is a vital option for saving low-entropy matter/energy for future generations.

23.3 Photovoltaic Power – Changing Our Energy Basis Toward Sustainability

After having derived the importance of tapping solar energy for economic processes, we shall now review the role of photovoltaic power technologies in a future energy market.

In the introduction of this doctoral thesis, we have been acquainted with Amory Lovin's illustrative statement of using high-quality energy forms for low-grade

---

[120] Georcescu-Roegen, Nicholas (1995). *Summary of the Entropy Law and the Economic Problem*, p. 179.

energetic purposes. Forms of "chainsaw-economic" energy uses are common throughout our current energy system. It is hard to eliminate wasteful energy uses in large integrated central grid systems providing reliable low-cost electricity. However, economic costs do not always represent real costs. Environmental and social costs are rarely being considered in economic consumer and business calculations and investment decision making.

Photovoltaic power generation is one of the most expensive forms of electricity production. Yet, it also demonstrates the high quality of electric power. In remote applications without grid-connectivity, energy production and consumption are much more inclined to take into account energetic arguments, resulting in initially cutting down energy needs as low as possible before providing electric power to only those appliances that cannot be operated with another energy form.

Electric consumers in grid-connected central power systems have lost this susceptibility of valuing the quality of energy used. Ubiquitous availability of low-cost electric power has led to a broad array of inefficient energy uses. Photovoltaic technologies are one form of renewable power generation that features low total emissions; high conversion efficiencies without any mechanical transformation component; and universal applicability at decentral locations near the customer load, minimizing systemwide electric losses. Increasing the total share of PV power generation in our electricity system would increase our consciousness of the real value of electric power and soon indicate that heat-electric-heat transformation processes are both energetically inefficient and environmentally absurd.

Small grid-connected residential PV systems could therefore not only shape a more decentralized electric power system, but also modify our energy system toward a more energy-efficient use of natural resources. A wide array of solar-based technologies as well as further efficiency improvements and substitutions of unfavorable energy sources are necessary to create a sustained power system for industrialized, transitional, and developing countries. Distributed renewable power generation helps to respect the regional environmental carrying capacities and is a crucial prerequisite to a sustainable global development.

PART X.

APPENDIX

24. Units of Measure

Currency Units		
DM	Exchange Rate 1 US$ = 1.80 DM	German Mark
DPf		German Pfennig
US$		US Dollar
US cent		US Cent
Electric Power Units		
W	Watt	Electric Power Generation Capacity
kW	Kilowatt	1000 Watts
MW	Megawatt	1000 Kilowatts
Wp	Watt Peak	Maximum Electric Power Generation Capacity
kWp	Kilowatt Peak	1000 Watts Peak
MWp	Megawatt Peak	1000 Kilowatts Peak
kWh	Kilowatthour	Amount of Electricity Consumed / Produced

25. Glossary

US English explanations and definitions are partly analogous to Ed Smeloff and
Peter Asmus (1997), *Reinventing Electric Utility Systems*, pp. 209-215.

Annual Fixed Charge (AFC)
  The present value of an investment converted
  to a stream of equal annual payments.
	Annuität

Avoided Costs (see Marginal Costs)
  Amount of money a utility would have to
  spend for the next increment of electric
  generation that it instead buys from a
  cogenerator or independent power producer.
	Vermeidungskosten

Blackout Electric Outage.	Stromausfall

Capacity
  Maximum electricity output of a power plant
  under specified conditions.
	Kapazität

Cogeneration
  Production of heat energy and electrical power
  at the power plant.
	Kraft-Wärme-Kopplung

Cost Recovery Factor (CRF)
  Factor to multiply with the investment cost in
  order to obtain the Annual Fixed Charge.
	Kapitalwiedergewinnungs-
	  faktor

Decentralized Power Generation
  Distributed electricity production with a
  multitude of small dispersed generators.
	Dezentrale Stromerzeugung

Decommissioning
  Dismantling of a power plant.
	Stillegung / Abbau eines
	  Kraftwerks

Demand Side Management (DSM)
  Includes energy efficiency efforts, load
  management, fuel substitution, etc.
	Nachfrageorientiertes
	  Energiemanagement

Deregulation
  Introduction of competition to the electric
  power industry.
	Deregulierung des
	  Strommarktes

Distributed (Power) Generation
  Decentralized electricity production with a
  multitude of small dispersed generators.
	Dezentrale Stromerzeugung

Distributed Resources
  Concept of decentral power, water, and other
  resource systems.
	Dezentrale Energie-, Wasser-
	  und Rohstoffversorgung

Distributed Utility
  Electric power company using a multitude of
  small dispersed generators to produce power.
	Dezentral ausgerichtetes
	  Stromversorgungs-
	  unternehmen

Distribution; Electric Distribution System Power lines to end-use customers.	Endverteilung von Energie
Electric Customers / Consumers	Stromkunden
Electricity; Electric Power; Power; Energy Synonyms for electricity.	(hier:) Elektrizität

Externalities
  Costs or benefits not accounted for in the
  market price.
	Externalitäten

Fossil Fuels Oil, natural gas, coal, and nuclear fuels.	Fossile Brennstoffe

Fuel Cell
  Device that converts chemical energy
  (hydrogen, methanol, natural gas) into
  electricity without burning it.
	Brennstoffzelle

Generation; (Electric) Power Generation Production of electric power.	Stromerzeugung

Grid; Central / Public Grid System The transmission and distribution system that links power plants to customers.	Öffentliches Stromnetz
Independent System Operator (ISO) Entity coordinating physical power demand and supply and supervising access to the central power grid.	Independent System Operator (Instanz zur Koordinierung von Stromangebot und -nachfrage im zentralen Netz)
Integrated Resource Planning (IRP) Least-cost planning method comparing energy efficiency efforts and demand side management alternatives to additions in generation capacity.	Integrierte Ressourcenplanung (Least-Cost-Planning)
Investor-Owned Utility Utility owned by shareholders.	Stromversorger in Form eines Kapitalunternehmens
Levelized Cost Calculation Calculation that converts the present value of an investment to a stream of equal annual payments.	Annuitätenrechnung
Liberalization Introduction of competition to the electric power industry.	Liberalisierung des Strommarktes
Load; Electric Load Electricity supplied to meet a user's demand.	Stromverbrauch
Losses Wasted energy.	Verluste
Marginal Costs (see Avoided Costs) Amount of money a utility would spend for producing an additional unit of electric power.	Grenzkosten
Micro-Grid Systems / Micro-Grids Small electrical power grids isolated from the central grid system.	Micro-Grid (kleine elektrisch isolierte Stromnetze)

Municipal Utility	Stadtwerk, kommunaler Stromversorger
Nominal Values Include the effects of inflation.	Nominale Werte
Outage Electric Blackout.	Stromausfall
Peak Load Highest electrical demand.	Stromspitzen
Photovoltaic Cell / Module Semiconductor that converts light directly into electric current.	Photovoltaikzelle / -modul
Power Exchange Institution that sets the current electricity price according to demand and supply bids.	Strombörse
Procurements (of PV components) Volumes of PV systems ordered and purchased	Käufe (von PV-Systemen, i.S.v. Einkaufsmengen)
Rates; Electricity Rates; Electric Rates Price of electricity per kWh.	Stromtarif, Stromkosten (je kWh)
Real Values Do not include the effects of inflation.	Reale (inflationsbereinigte) Werte
Renewables, Renewable Energies Inexhaustible energies such as solar radiation, wind and hydropower.	Erneuerbare Energien
Residential Customers / Consumers	Private Stromkunden; Tarifkunden
Restructuring Introduction of competition to the electric power industry.	Deregulierung des Strommarktes

Retail Market Sale of utility electricity to industrial or residential customers.	Endverkaufsmarkt
Service Provider Electricity firm in a deregulated environment.	Energiedienstleistungsunter- nehmen
Small Power Producers (SPPs) E.g. Residential power producers / generators	Kleine Stromerzeuger (z.B. private Konsumenten)
Substation Facility that steps up or steps down the voltage in utility power lines.	Umspannwerk, Transformator
Surplus Energy Excess electric power that a small power producer does not use himself and feeds into the grid.	Überschußstrom
Tailored / Interruptible Service Electric contract that allows the supplier to stop services at times.	Stromvertrag mit vereinbarten Sperrzeiten
Transmission; Electric Transmission System Power lines for long-distance power transport.	Elektrizitäts(fern)über- tragung
Turbine Generator Transforms any kind of flow energy into electricity.	Elektrische Turbine / Generator
Unbundling Institutional Splitting of electric power production, transmission, and distribution services	Unbundling (Institutionelle Aufteilung von Stromer- zeugung, Übertragung und Verteilung)
Utilities; (Electric) Utilities (Electric) power companies.	Stromversorgungsunter- nehmen

Vertical Integration
    Control or ownership of subsequent services
    by a single company (e.g. transmission,
    generation, and distribution).
	Vertikale Integration

Wholesale Market Trade of electric power between utilities.	Strom(groß)handel

26. References

Balzhiser, Richard E. (1996). Technology – It's Only Begun to Make a Difference. *The Electricity Journal*, May 1996, USA.

Bayernwerk, Siemens, RWE (April 1993). *Kostenentwicklung von Photovoltaik-Kraftwerken in Mitteleuropa*. Bayernwerk, München, Germany.

Bayless, Charles E. (1994). Less is More: Why Gas Turbines Will Transform Electric Utilities. *Public Utilities Fortnightly*, USA, December 1, 1994.

Beyer, Ulrich (1998). Konzeption und Kundenerfahrung mit dem Umwelttarif der RWE Energie AG. In *Proceedings of the 5th Congress RENERGIE 98*, Hamm, Nordrheinwestfalen, Germany, June 5-6, 1998. Internationales Wirtschaftsforum Regenerative Energien (Ed.), Münster, Germany.

Bischof, Ralf (1998). Erneuerbare Energien unterm Deckel. *Photon*, Jan/Feb 1998. Anne Kreuzmann (Ed.), Solar Verlag, Aachen, Germany.

Brennan, Timothy J., Karen L. Palmer, et al. (1996). *A Shock to the System. Restructuring America's Electricity Industry*. Resources for the Future, Washington D.C., USA.

Bullinger, Hans-Jörg (1998). Mega-Konzerne bald mächtiger als Staaten. *Nürtinger Zeitung*, German Daily Newspaper, June 13, 1998. Nürtingen, Germany.

Bundesministerium für Umwelt, Naturschutz und Reaktorsicherheit BMU (1992). *Zur Novellierung des Energiewirtschaftsgesetzes – Defizitanalyse und Reformkonzeption aus umweltpolitischer Sicht*. BMU, Bonn, Germany.

Byrnes, Brian et al. (1996). Green Pricing: The Bigger Picture. *Public Utilities Fortnightly*, August 1996, USA.

Bzura, John J. (1990). *The New England Electric Photovoltaic Systems Research and Demonstration Project*. New England Electric Services, Westborough, MA, USA.

Bzura, John J. (1994). *Photovoltaic Research and Demonstration Activities at New England Electric*. IEEE Power Engineering Society and New England Electric Services, Westborough, MA, USA.

Bzura, John J. (1997). *Basic Economics of Residential PV Systems*, 1997 Update. New England Electric Services, Westborough, MA, USA.

Daly, Herman E. (1996). *Beyond Growth. The Economics of Sustainable Development.* Beacon Press, Boston/MA, USA, 1996.

Deutsche Gesellschaft für Sonnenenergie (1998), *Förderung Thermischer Solaranlagen und Photovoltaik-Anlagen.* Information Brochure January 1998. Deutsche Gesellschaft für Sonnenenergie, München, Germany.

Dimmler, Bernhard, E. Gross et al. (1997). Progress in CIGS Large Area Thin Film PV Modules Based on Industrial Process Technology. Zentrum für Sonnenenergie- und Wasserstoff-Forschung (ZSW), Stuttgart, Germany. Paper presented to the *14th PV Solar Energy Conference*, Barcelona, Spain, June 30th to July 4th, 1997.

Edinger, Raphael (1998). Electric Restructuring and Photovoltaic Technologies - Strategies for a Deregulated Power Market. Paper presented to the *Boston Consulting Group Energy & Utilities Meeting*, Hotel Treudelberg, Hamburg, Germany, February 5th, 1998.

Edinger, Raphael (1998). Perspektiven der PV-Marktentwicklung in Deutschland und den USA. In *Proceedings of the 5th Congress RENERGIE 98*, Hamm, Nordrheinwestfalen, Germany, June 5-6, 1998. Internationales Wirtschaftsforum Regenerative Energien (Ed.), Münster, Germany.

Edinger, Raphael (1998). Wirtschaftliche Aspekte von Solarstrom in einem liberalen Elektrizitätsmarkt. Eine ökonomische Analyse netzgebundener Photovoltaikanlagen. Paper presented to the *EUROSOLAR Conference "Der Öko-Strommarkt"*, at Haus der Wirtschaft, Stuttgart, Germany, April 25th, 1998.

Edinger, Raphael and Sanjay Kaul (1999). *Renewable Resources for Electric Power. Prospects and Challenges.* Quorum Books, Westport, USA. In press.

Energy Information Administration EIA / US Department of Energy DOE (1996). *The Changing Structure of the Electric Power Industry: An Update.* EIA / DOE, Washington, D.C., USA.

Energy Information Administration EIA / US Department of Energy DOE (1993). *The Changing Structure of the Electric Power Industry, 1970-1991.* EIA / DOE, Washington, D.C., USA.

Energy Information Administration EIA / US Department of Energy DOE (1995), *Electricity Generation and Environmental Externalities: Case Studies.* EIA / DOE, Washington, D.C., USA.

Energy Information Administration EIA / US Department of Energy DOE (1997), *Annual Energy Outlook 1998. With Projections to 2020.* EIA / DOE, Washington, D.C., USA.

Energy Information Administration EIA / US Department of Energy DOE (1995). *Renewable Energy Annual 1995.* EIA / DOE, Washington, D.C., USA.

Energy Information Administration EIA / US Department of Energy DOE (1997). *Renewable Energy Annual 1996.* EIA / DOE, Washington, D.C., USA.

Energy Information Administration EIA / US Department of Energy DOE (1998). Industry Developments: California Successfully Starts Deregulated Electricity Market. *Electric Power Monthly – Industry Developments,* May 1998. EIA / DOE, Washington, D.C., USA.

Fox-Penner, Peter (1997). *Electric Utility Restructuring. A Guide to the Competitive Era.* Public Utilities Reports, Inc., Vienna, Virginia, USA.

Fraunhofer Institut für Solare Energiesysteme (ISE), *1000-Dächer Meß- und Auswerteprogramm, Jahresjournal 1995.* ISE, Freiburg, Germany.

Genennig, Bernd, and Volker U. Hoffmann (1996). *Sozialwissenschaftliche Begleituntersuchung zum Bund-Länder-1000-Dächer Photovoltaik-Programm.* Umweltinsitut Leipzig and Fraunhofer Institut für Solare Energiesysteme, Leipzig, Germany.

Georcescu-Roegen, Nicholas (1995). Summary of the Entropy Law and the Economic Problem. In: *A Survey of Ecological Economics.* Krishnan, Harris, and Goodwin (Eds.). Island Press, Washington D.C. / Covelo, CA, USA, 1995.

Gronbeck, Christopher (1998). *PV Cell Basics.* URL http://Solstice.crest.org/renewables/re-kiosk/solar/pv/theory/physics/basics.htm.

Gronbeck, Christopher (1998). *Trackers and Concentrators.* URL http://Solstice.crest.org/renewables/re-kiosk/solar/pv/theory/track/index.shtml.

Hagenmeyer, Ernst (1997). Neue Märkte und neue Player – Strukturwandel in der deutschen Stromwirtschaft. In *Entscheidungsfindung und –steuerung bei wachsender Unsicherheit in der Energiewirtschaft*. Energiewirtschaftliches Institut an der Universität Köln (Ed.), München, Germany.

Hamburger Electricitäts-Werke HEW AG (1995). *Ausbau der Photovoltaik in Hamburg. Technik, Wirtschaftlichkeit und Fördermodelle*. HEW, Hamburg, Germany.

Hoff, Thomas E. and the Pacific Energy Group (1997), *Integrating Renewable Energy Technologies in the Electric Supply Industry: A Risk Management Approach*. Report for the National Renewable Energy Laboratory (NREL), Golden, CO, USA.

Hoff, Thomas E., Howard J. Wenger, and Brian K. Farmer (1996). Distributed Generation: An Alternative to Electric Utility Investments in System Capacity. *Energy Policy* 24(2), 1996, USA.

Hoff, Thomas E., Howard J. Wenger, Christy Herig, and Robert W. Shaw. (1998). A Micro-Grid With PV, Fuel Cells, and Energy Efficiency. In *Proceedings of the 1998 Annual ASES Conference*, Albuquerque, NM, June 1998. American Solar Energy Society (Ed.), Boulder, CO, USA.

Holt, Edward A. (1997). Green Pricing Experience and Lessons Learned. In *Proceedings of the 1996 ACEEE Summer Study*, Pacific Grove, California, August 25-31, 1996. American Council for an Energy-Efficient Economy (Ed.), Washington, D.C., USA.

Howard, Milton R. (PanEnergy, 1996). Advancing Electric Competition by Providing Electric Power Choice. Paper presented to the *17th Annual North American Conference of the United States Association for Energy Economics / International Association for Energy Economics, "(De) Regulation of Energy: Intersecting Business, Economics and Policy"*. Boston Park Plaza Hotel, October 27-30, 1996, USA.

Hyman, Leonard S. (1997). *America's Electric Utilities. Past, Present and Future*. 6th Edition. Public Utilities Reports, Inc., Arlington, Virginia, USA.

Informationszentrale der Elektrizitätswirtschaft e.V. (1997). Liberalisierung des Strommarktes: Koalition bei Energierecht einig. *Stromthemen* #12, December 1997. Frankfurt/Main, Germany.

Informationszentrale der Elektrizitätswirtschaft e.V. (1998). Umstrukturierung für den Wettbewerb. Bereiche Erzeugung, Transport und Verteilung werden getrennt. *Stromthemen* #4, April 1998. Frankfurt/Main, Germany.

Informationszentrale der Elektrizitätswirtschaft e.V. (1998). Verbändevereinbarung über Stromdurchleitung. Durchbruch für Wettbewerb. *Stromthemen* #5, May 1998. Frankfurt/Main, Germany.

Internationales Wirtschaftsforum Regenerative Energien IWR (1998). *Förderung/Energieberatung in Baden-Württemberg.* URL http://www.uni-muenster.de/Energie/ Münster University, Münster, Germany.

Internationales Wirtschaftsforum Regenerative Energien IWR (1998). *Vergütungssätze für Strom aus erneuerbaren Energien.* URL http://www.uni-muenster.de/Energie/ Münster University, Münster, Germany.

Kreutzmann, Anne (1997). Phönix fliegt für Photovoltaik. *Photon,* September/October 1997. Anne Kreuzmann (Ed.), Solar Verlag, Aachen, Germany.

Lamarre, Leslie (1997). Utility Customers Go for the Green. *EPRI Journal,* March/April 1997, USA.

Langniß, Luther, Nitsch and Wiemken (1997). *Strategien für eine nachhaltige Energieversorgung – Ein solares Langfristszenario für Deutschland.* Deutsches Zentrum für Luft- und Raumfahrt e.V. (DLR) and Fraunhofer Institut für Solare Energiesysteme (ISE), Stuttgart / Freiburg, Germany.

Letendre, Steven, John Byrne, and Young-Doo Wang (1996). The Distributed Utility Concept: Toward a Sustainable Electric Utility Sector. In *Proceedings of the 1996 ACEEE Summer Study,* Vol. 7, Asilomar, CA. American Council for an Energy-Efficient Economy, Washington D.C., USA.

Libby, Leslie (1997). Austin's Solar Explorer Program. In *Proceedings of the ASES Solar 97 Conference,* Washington, DC, April 1997. American Solar Energy Society (Ed.), Boulder, CO, USA.

Lovins, Amory B. (1977). *Soft Energy Paths. Toward a Durable Peace.* Harper & Row, Publishers, Inc., New York, USA.

Maycock, Paul D. (1997). *Photovoltaic Technology, Performance, Cost and Market Forecast: 1975 – 2010.* Executive Summary to the Sixth Edition, May 1997. PV Energy Systems, Warrenton VA, USA.

Maycock, Paul D. (1997). *World Price for Photovoltaic Modules, 1975-95.* Database Diskette January 1997, File Solar.wk1. Worldwatch Institute (Ed.), Washington, D.C., USA.

Mez, Lutz (1997). The German Electricity Reform Attempts: Reforming Co-optive Networks. In *European Electricity Systems in Transition.* Atle Midttun (Ed.), Elsevier Science, Oxford, UK.

Molly, Jens Peter, and Knud Rehfeldt (1997). Wirtschaftlichkeit von Windanlagen / Aktuelle Kostenentwicklung. In: *Allgemeine Entwicklung der Kosten der Windstromerzeugung in Deutschland. Studie im Auftrag des Bundesverbandes Windenergie e.V.(BWE),* March 1997, Part 2. Deutsches Windenergie-Institut (DEWI), Wilhelmshaven, Germany.

Monopolkommission (1976). *Mehr Wettbewerb ist möglich.* Hauptgutachten 1973/75. Nomos, Baden-Baden, Germany.

Müller, Jürgen and Konrad Stahl (1996). Regulation of the Market for Electricity in the Federal Republic of Germany. In: *International Comparisons of Electricity Regulation.* Richard J. Gilbert and Edward P. Kahn (Eds.), Cambridge University Press, USA.

Oak Ridge National Laboratory (1994). *The Impact of Environmental Externality Requirements on Renewable Energy,* Table A-1. Unpublished Report, prepared for the Energy Information Administration / US Department of Energy. Oak Ridge, Tennessee, July 1994.

Oberländer, Hans-Ulrich (1998). Alternative Förderkonzepte für Photovoltaik-Anlagen. In *Proceedings to the 13th Symposium Photovoltaic Solar Energy.* Ostbayerisches Technologie Transfer Institut OTTI e.V. (Ed.), Regensburg, Germany.

Osborn, Donald E. (1997). Commercialization of Utility PV Distributed Power Systems. In *Proceedings of the ASES Solar 97 Conference,* Washington, DC, April 1997. American Solar Energy Society (Ed.), Boulder, CO, USA.

Pontenagel, Irm (1998). Präsentation von Label-Ideen für exclusive Öko-Stromanbieter und Öko-Stromkunden. Presentation to the *EUROSOLAR Conference "Der Öko-Strommarkt"*, at Haus der Wirtschaft, Stuttgart, Germany, April 25th, 1998.

Räuber, Armin (1998). Weltweite PV-Aktivitäten – eine kritische Bewertung. In *Proceedings to the 13th Symposium Photovoltaic Solar Energy*. Ostbayerisches Technologie Transfer Institut OTTI e.V. (Ed.), Regensburg, Germany.

Rocky Mountain Institute (1997). Tunneling Through the Cost Barrier. In *RMI Newsletter*, Vol. XIII, Number 2, Summer 1997. Rocky Mountain Institute, Snowmass, CO, USA.

Ruth, Matthias and Bruce Hannon (1997). *Modeling Dynamic Economic Systems.* Springer Verlag, New York, USA.

Schulz, Manfred (1998). Marktperspektiven und ordnungspolitischer Handlungsrahmen für regenerative Energien. In *Proceedings of the 5th Congress RENERGIE 98*, Hamm, Nordrheinwestfalen, Germany, June 5-6, 1998. Internationales Wirtschaftsforum Regenerative Energien (Ed.), Münster, Germany.

Seltmann, Thomas (1995). Brauchen wir Einkaufsgemeinschaften? *Solarbrief* #2/95, Solarenergie-Förderverein (Ed.), Aachen, Germany.

Siegele, Ludwig (1998). Es klingelt bei den Kartellwächtern. Fusionen zwischen den amerikanischen Telphongesellschaften bedrohen den freien Wettbewerb. *DIE ZEIT* #22, May 20th, 1998, Germany.

Smeloff, Ed, and Peter Asmus (1997). *Reinventing Electric Utilities. Competition, Citizen Action, and Clean Power.* Island Press, Washington, DC / Covelo, CA, USA.

Solarenergie-Förderverein (1996). Die Kostendeckende Vergütung. *Solarbrief* #4/96, Solarenergie-Förderverein (Ed.), Aachen, Germany.

Solarenergie-Förderverein (1997). Das Aachener Modell. *Solarbrief* #1/97, Solarenergie-Förderverein (Ed.), Aachen, Germany.

Solarenergie-Förderverein (1997). Erfolge der kostendeckenden Vergütung. *Solarbrief* #3/97, Solarenergie-Förderverein (Ed.), Aachen, Germany.

Springorum, Roland, and Anne Kreuzmann (1996). Solardachziegel. Strom aus dem Dach. *Photon,* May/June 1996. Anne Kreuzmann (Ed.), Solar Verlag, Aachen, Germany.

Teske, Sven (1998). Die Cyrus-Solarkampagne von Greenpeace geht in das dritte Jahr. In *Proceedings to the 13th Symposium Photovoltaic Solar Energy.* Ostbayerisches Technologie Transfer Institut OTTI e.V. (Ed.), Regensburg, Germany.

The Electricity Journal (1997). RDI Sees 20 Utilities Holding $100 Billion in Stranded Costs. *The Electricity Journal,* Vol. 10, #3, April 1997. Marritz, Robert O. (Ed.), Seattle, WA, USA.

The Results Center (1994). *Sacramento Municipal Utility District Solar Photovoltaic Program, Profile #111.* The Results Center IRT Environment, Inc., Aspen/Colorado, USA.

Utility PhotoVoltaic Group (UPVG), Multi-Utility Market Survey on "PV Friendly" Pricing Reveals Strong, Positive Reaction to PV. *The UPVG Record,* Fall 1996. Utility PhotoVoltaic Group (Ed.), Washington D.C., USA.

Vereinigung Deutscher Elektrizitätswerke VDEW (1997). *Jahreserhebung bei Elektrizitätsversorgungsunternehmen für das Jahr 1996. Tabelle 1.1 Stromabsatz und Erlöse.* VDEW, Frankfurt/M., Germany.

Vereinigung Deutscher Elektrizitätswerke VDEW (1997). *Strom-Daten Februar 1997. Kapitel 12: Brennstoff- und Erzeugungskosten.* VDEW, Frankfurt/M., Germany.

Vereinigung Deutscher Elektrizitätswerke VDEW (1998). *Drei-Personen-Haushalt Januar 1998: 91 Mark für Strom.* VDEW News Internet Publication, Germany, URL http://www.strom.de/

Vesey, Andrew (1997). 2,000 by 2,000: A Challenge. *PV Vision,* Volume 5, #1, Winter 1997. Utility PhotoVoltaic Group UPVG, Washington, D.C., USA.

Vogl, Reiner J., Manfred M. Gößl, and Gerhard M. Feldmeier (1997). *Die Elektrizitätswirtschaft in der Bundesrepublik Deutschland. Wettbewerbsstruktur im Kontext europäischer Energiepolitik.* Hänsel-Hohenhausen, Egelsbach / Frankfurt / St. Peter Port, Germany.

Weizsäcker, Ernst Ulrich von, Amory B. Lovins, and L. Hunter Lovins (1996). *Faktor Vier. Doppelter Wohlstand – halbierter Naturverbrauch.* Droemer Knaur, München, Germany.

Wenger, Howard J., Thomas E. Hoff, and Donald E. Osborn (1997). A Case Study of Utility PV Economics. In *Proceedings of the ASES Solar 97 Conference*, Washington, DC, April 1997. American Solar Energy Society (Ed.), Boulder, CO, USA.

Wenger, Howard J., Thomas E. Hoff, and Jan Pepper (1996). *Photovoltaic Economics and Markets: The Sacramento Municipal Utility District as a Case Study.* California Energy Commission, California, USA.

Wenger, Howard, Christy Herig, et al. (1996). Niche Markets for Grid-Connected Photovoltaics. Paper presented to the *IEEE Photovoltaic Specialists Conference* Washington, D.C., USA. May 13-17, 1996.

Wiesner, Wolfgang, TÜV Rheinland (1998). Zertifizierungsgrundlage für Anbieter und Kunden des Öko-Strommarktes. Presentation to *the EUROSOLAR Conference "Der Öko-Strommarkt"*, at Haus der Wirtschaft, Stuttgart, Germany, April 25th, 1998.

Worldwatch Institute (1997). *Government Research and Development Spending in International Energy Agency Member Countries.* Database Diskette January 1997, File R&D.wk1. Worldwatch Institute, Washington, D.C., USA.

27. Index

27.1 Index of Figures

Figure 3-1. and Figure 3-2. Edinger, Raphael and Sanjay Kaul (1999). *Renewable Resources for Electric Power. Prospects and Challenges.* Quorum Books, Westport/USA. In press.

Figure 3-3. Corrected Chart Version; Original from Mez, Lutz (1997). *The German Electricity Reform Attempts: Reforming Co-optive Networks*, p. 233.

Figure 4-1. Gronbeck, Christopher (1998). *PV Cell Basics.* URL http://Solstice.crest.org/renewables/re-kiosk/solar/pv/theory/physics/basics.htm.

Figure 4-3. Gronbeck, Christopher (1998). *Trackers and Concentrators.* URL http://Solstice.crest.org/renewables/re-kiosk/solar/pv/theory/track/index.shtml.

Figure 5-1. Maycock, Paul D. (1997). *World Price for Photovoltaic Modules, 1975-95.* Worldwatch Institute Database Diskette January 1997, File Solar.wk1.

Figure 5-2. and Figure 5-3. Wenger, Howard, Christy Herig, et al. (1996). *Niche Markets for Grid-Connected Photovoltaics*, p. 3.

Figure 5-4. Data from Dimmler, Bernhard, E. Gross et al. (1997). *Progress in CIGS Large Area Thin Film PV Modules Based on Industrial Process Technology*, p. 4.

Figure 5-5. Data from Solarenergie-Förderverein (1997). *Erfolge der kostendeckenden Vergütung*, p. 25.

Figure 6-2. Edinger, Raphael (1998). *Perspektiven der PV-Marktentwicklung in Deutschland und den USA*, p. 179.

Figure 6-3. Data from Beyer, Ulrich (1998). *Konzeption und Kundenerfahrung mit dem Umwelttarif der RWE Energie AG*, p. 27.

Figure 6-4. Data from Wenger, Howard J., Thomas E. Hoff, and Jan Pepper (1996). *Photovoltaic Economics and Markets: The Sacramento Municipal Utility District as a Case Study*, p. 5-3.

Figure 6-5. Wenger, Howard J., Thomas E. Hoff, and Donald E. Osborn (1997). *A Case Study of Utility PV Economics*, p. 175.

Figure 7-1. Bayless, Charles E. (1994). *Less is More: Why Gas Turbines Will Transform Electric Utilities*, p. 24.

Figure 7-2. and Figure 7-3. Hoff, Thomas E., Howard J. Wenger, and Brian K. Farmer (1996). *Distributed Generation: An Alternative to Electric Utility Investments in System Capacity*, pp. 137/147.

Figure 7-6. Wenger, Howard J., Thomas E. Hoff, and Jan Pepper (1996). *Photovoltaic Economics and Markets: The Sacramento Municipal Utility District as a Case Study*, p. 4-11.

Figure 7-7. Hoff, Thomas E., Howard J. Wenger, Christy Herig, and Robert W. Shaw. (1998). *A Micro-Grid With PV, Fuel Cells, and Energy Efficiency*, p. 2.

Figure 8-12. Energy Information Administration EIA / US Department of Energy DOE (1996). *The Changing Structure of the Electric Power Industry: An Update*, p. 36.

Figure 9-1. Daly, Herman E. (1996). *Beyond Growth. The Economics of Sustainable Development*, p. 29.

27.2 Index of Tables

Table 4-1., Table 4-2., and Table 4-3. Data from Maycock, Paul D. (1997). *Photovoltaic Technology, Performance, Cost and Market Forecast: 1975 – 2010*, pp. 12-13.

Table 5-1. Data from Wenger, Howard, Christy Herig, et al. (1996). *Niche Markets for Grid-Connected Photovoltaics*, p. 4.

Table 5-2. Edinger, Raphael (1998). *Wirtschaftliche Aspekte von Solarstrom in einem liberalen Elektrizitätsmarkt. Eine ökonomische Analyse netzgebundener Photovoltaikanlagen*, p. 2.

Table 5-3. Data from Vereinigung Deutscher Elektrizitätswerke VDEW (1997). *Strom-Daten Februar 1997. Kapitel 12: Brennstoff- und Erzeugungskosten*.

Table 5-4. Data compiled from
(a) Fox-Penner, Peter (1997). *Electric Utility Restructuring. A Guide to the Competitive Era*, p. 387;
(b) The Electricity Journal, USA (April 1997). *RDI Sees 20 Utilities Holding $100 Billion in Stranded Costs*, p. 4.

Table 5-5. Data from Worldwatch Institute (1997). *Government Research and Development Spending in International Energy Agency Member Countries.* Database Diskette January 1997, File R&D.wk1.

Table 5-6. Data from Oak Ridge National Laboratory (1994). *The Impact of Environmental Externality Requirements on Renewable Energy,* Table A-1.

Table 5-7. Data from Internationales Wirtschaftsforum Regenerative Energien IWR (1998). *Vergütungssätze für Strom aus erneuerbaren Energien.* URL http://www.uni-muenster.de/Energie/

Table 6-1. Data from Wenger, Howard J., Thomas E. Hoff, and Jan Pepper (1996). *Photovoltaic Economics and Markets: The Sacramento Municipal Utility District as a Case Study,* p. 5-3.

Table 8-3. Data from:
– Fraunhofer Institut für Solare Energiesysteme (ISE), *1000-Dächer Meß- und Auswerteprogramm, Jahresjournal 1995,* p. 6.
– Bzura, John J. (1997). *Basic Economics of Residential PV Systems,* p. 1.
– Wenger, Howard J. *PV Cost Calculations,* Personal Email Conversation, July 18th, 1997. Sacramento Municipal Utility District, Sacramento, CA, USA.

Table 8-7. Data from Vereinigung Deutscher Elektrizitätswerke VDEW (1997). *Jahreserhebung bei Elektrizitätsversorgungsunternehmen für das Jahr 1996. Tabelle 1.1 Stromabsatz und Erlöse.*

Table 8-12. Data from Springorum, Roland, and Anne Kreuzmann (1996). *Solardachziegel. Strom aus dem Dach,* p. 13.

Table 8-13. Data from Dimmler, Bernhard, E. Gross et al. (1997). *Progress in CIGS Large Area Thin Film PV Modules Based on Industrial Process Technology,* p. 4.

Table 8-17. Data from Wenger, Howard J., Thomas E. Hoff, and Jan Pepper (1996). *Photovoltaic Economics and Markets: The Sacramento Municipal Utility District as a Case Study,* pp. ii and 4-6.

Table 8-19. Data from Hoff, Thomas E., Howard J. Wenger, Christy Herig, and Robert W. Shaw. (1998). *A Micro-Grid With PV, Fuel Cells, and Energy Efficiency,* pp. 1-6.

www.ingramcontent.com/pod-product-compliance
Lightning Source LLC
Chambersburg PA
CBHW031933220326
41598CB00062BA/1888